STEPPING MOTORS:
a guide to modern theory and practice

P. P. Acarnley

PETER PEREGRINUS LTD
on behalf of the
Institution of Electrical Engineers

Published by: The Institution of Electrical Engineers, London
and New York
Peter Peregrinus Ltd., Stevenage, UK, and New York
© 1982: Peter Peregrinus Ltd.

British Library Cataloguing in Publication Data

Acarnley, P.P.
Stepping motors. — (IEE control engineering
series; no. 19)
1. Stepping motors
I. Title   II. Series
621.46′2   TK2511

ISBN 0-906048-75-3

Printed in England by Short Run Press Ltd., Exeter

Contents

Preface

Stepping motors have been available for many years, but commercial exploitation only began in the 1960's when improved transistor fabrication techniques made available devices capable of switching large d.c. currents in the motor windings. This feature of switched winding currents gives the stepping motor its unique 'digital machine' properties, which are a considerable asset when interfacing to other digital systems. The rapid growth of digital electronics throughout the 1970's assured the stepping motor's future and today there is worldwide interest in its manufacture and application.

The many recent developments in stepping motor technology are documented in the excellent publications associated with regular conferences held at the University of Illinois, USA and the University of Leeds, UK. However, these sources of information are not generally suitable for the non-specialist and therefore the primary aim of this book is to provide an introductory text, which will enable the reader to appreciate the essential characteristics of stepping motors and the ways in which these characteristics are exploited in new motors, drives and controllers. No previous experience of electrical machines is assumed, but a general education in electronic engineering to HNC/first-year degree level would be a useful background, because some secondary topics (e.g. transistor switching, logic circuit design and microprocessor programming) receive only the briefest treatment.

For industrial users of stepping motors the book provides a basic theoretical approach to the more significant aspects of performance and relates this theory to recent developments. For example, the advantages of bilevel and chopper drive circuits are readily appreciated when it is realised that drive supply voltage is the most dominant influence on pull-out torque/speed characteristics. Armed with a sound theoretical understanding of stepping motor operation, the reader will also be in a position to assess the possibilities of new innovations. One important area of development is the microprocessor-based stepping motor control, so the final Chapter reviews the state-of-the-art and indicates where future advances are likely to occur.

Considerable interest has been shown in the possibilities of the stepping motor as an educational tool (Turner, 1978; Rooy *et al*, 1979; Barnard and Lloyd, 1980); the motor can be easily interfaced to a microprocessor, which is then able to control a device doing mechanical work. A growing number of institutions include the stepping motor on the syllabus and the theoretical sections of the book provide material suitable for lecture and examination purposes, while the later Chapters include topics suitable for project investigation. Regrettably several academically-satisfying subjects, notably optimisation of tooth profiles, have had to be omitted, because they are more relevant to the motor designer, who will already be conversant with the appropriate literature.

Finally I would like to thank my wife Rita and son David for their support and to thank Ward Crawford, John Halfpenny and Richard Danbury for their comments on the original draft.

P.P. Acarnley

Chapter 1

Stepping motors

1.1 Introduction

Since the earliest days of stepping motor development a large number of designs
have appeared, but it was soon realised that if the motor is to produce a significant
torque from a reasonable volume then both the stationary and rotating components
must have large numbers of iron teeth, which must be able to carry a substantial
magnetic flux. Surprisingly few methods of achieving this basic necessity have
withstood the test of commercial success and today we need only seriously consider
two types of stepping motor.

Although there is a wide range of stepping motor designs (Fig.1.1), most motors
can be identified as variations on the two basic types: variable-reluctance or hybrid.
Even if the internal geometry of a motor appears very different, it is always possible
to characterise its terminal behaviour in terms of one of these types and we are
therefore justified in concentrating attention on them in this Chapter and, indeed,
throughout the book. The reader is first introduced to the simple stepping action of
the variable-reluctance and hybrid types before going on, in later Chapters, to
consider how the motor can be integrated into a complete system.

The essential property of the stepping motor is its ability to translate switched
excitation changes into precisely defined increments of rotor position ('steps').
Accurate positioning of the rotor is generally achieved by magnetic alignment of
the iron teeth on the stationary and rotating parts of the motor. For the hybrid
motor the main source of magnetic flux is a permanent magnet and d.c. currents
flowing in one or more windings direct the flux along alternative paths. There are
two configurations for the variable-reluctance stepping motor, but in both cases
the magnetic field is produced solely by the winding currents.

1.2 Multi-stack variable-reluctance stepping motors

1.2.1 Principles of operation

The multi-stack variable-reluctance stepping motor is divided along its axial length

Fig. 1.1   *The range of stepping motors produced by one manufacturer*
[photograph by Warner Electric Inc., USA]

into magnetically isolated sections ('stacks'), each of which can be excited by a separate winding ('phase'). In the cutaway view of Fig.1.2, for example, the motor has three stacks and three phases, but motors with up to seven stacks and phases are available.

Each stack includes a stationary element ('stator'), held in position by the outer casing of the motor, and a rotating element. The rotor elements are fabricated as a

Fig. 1.2   *Cutaway view of a three-stack variable-reluctance stepping motor*
[photograph by Warner Electric Inc., USA]

single unit ('rotor'), which is supported at each end of the machine by bearings and includes a projecting shaft for the connection of external loads, as shown in Fig.1.3(a). Both stator and rotor are constructed from iron, which is usually laminated so that magnetic fields within the motor can change rapidly without causing excessive eddy current losses.

The stator of each stack has a number of poles – Fig.1.3(b) shows an example with four poles – and a part of the phase winding is wound around each pole to produce a radial magnetic field in the pole. Adjacent poles are wound in the opposite sense, so that the radial magnetic fields in adjacent poles are in opposite directions. The complete magnetic circuit for each stack is from one stator pole, across the air-gap into the rotor, through the rotor, across the air-gap into an adjacent pole, through this pole, returning to the original pole via the back-iron. This magnetic circuit is repeated for each pair of adjacent poles and therefore in the example of Fig.1.3(b) there are four main flux paths.

The position of the rotor relative to the stator in a particular stack is accurately defined whenever the phase winding is excited. Positional accuracy is achieved by means of the equal numbers of teeth on the stator and rotor, which tend to align so as to reduce the reluctance of the stack magnetic circuit. In the position where the stator and rotor teeth are fully aligned the circuit reluctance is minimised and the magnetic flux in the stack is at its maximum value.

The stepping motor shown in Fig.1.3(b) has eight stator/rotor teeth and is in the position corresponding to excitation of stack A. Looking along the axial length of the motor the rotor teeth in each stack are aligned, whereas the stator teeth have different relative orientations between stacks, so in stacks B and C the stator and rotor teeth are not fully aligned. The effect of changing the excitation from stack A to stack B is to produce alignment of the stator and rotor teeth in stack B. This new alignment is made possible by a movement of the rotor in the clockwise direction; the motor moves one 'step' as a result of the excitation change.

Another step in the clockwise direction can be produced by removing the excitation of stack B and exciting stack C. The final step of the sequence is to return the excitation to stack A. Again the stator and rotor teeth in stack A are

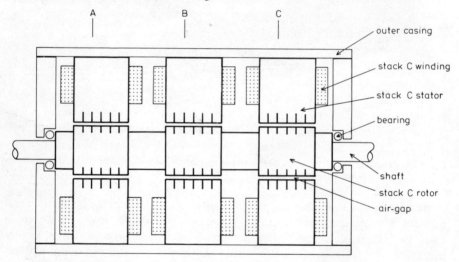

Fig. 1.3 (a) *Cross-section of a three-stack variable-reluctance stepping motor parallel to the shaft*

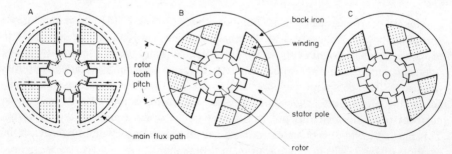

Fig. 1.3 (b) *Cross-sections of a three-stack variable-reluctance stepping motor perpendicular to the shaft*

fully aligned, except that the rotor has moved one rotor tooth pitch, i.e. the angle between adjacent rotor teeth defined in Fig.1.3(*b*). Therefore in this three-stack motor three changes of excitation cause a rotor movement of three steps or one rotor tooth pitch.

Continuous clockwise rotation can be produced by repeating the excitation sequence: *A, B, C, A, B, C, A, ...* Alternatively anticlockwise rotation results from the sequence: *A, C, B, A, C, B, A, ...* If bi-directional operation is required from a multi-stack motor it must have at least three stacks so that two distinct excitation sequences are available.

There is a simple relationship between the numbers of stator/rotor teeth, number of stacks and the step length for a multi-stack variable-reluctance motor. If the motor has *N* stacks (and phases) the basic excitation sequence consists of each stack being excited in turn, producing a total rotor movement of *N* steps. The same stack is excited at the beginning and end of the sequence and if the stator and rotor teeth are aligned in this stack the rotor must have moved one tooth pitch. Since one tooth pitch is equal to $360/p$ degrees, where *p* is the number of rotor teeth, the distance moved for one change of excitation must be:

$$\text{Step length} = 360/Np \text{ degrees} \qquad (1.1)$$

The motor illustrated in Fig.1.3 has three stacks and eight rotor teeth, so the step length is 15 degrees. For the multi-stack variable-reluctance stepping motor typical step lengths are in the range 2–15 degrees.

Successful multi-stack designs are often produced with additional stacks, so that the user has a choice of step length, e.g. a three-stack, sixteen rotor tooth motor gives a step of 7·5 degrees and by introducing an extra stack (with consequent reorientation of the other stacks) a 5·625 degree step is available. Although the use of higher stack numbers is a great convenience to the manufacturer, it must be remembered that more phase windings require more drive circuits, so the user has to pay a penalty in terms of drive circuit cost. Furthermore it can be shown (Acarnley, *et al*, 1979) that motors with higher stack numbers have no real performance advantages over a three-stack motor.

1.2.2   Aspects of design

Each pole of the multi-stack stepping motor is provided with a winding which produces a radial magnetic field in the pole when excited by a d.c. current. The performance of the stepping motor depends on the strength of this magnetic field; a high value of flux leads to a high torque retaining the motor at its step position. This relationship between torque and field strength receives more discussion in Chapter 3, so for the present we need only consider how the pole magnetic field can be maximised.

In the position where rotor and stator teeth are fully aligned, as in stack *A* of Fig.1.3(*b*), the reluctance of the main flux path is at its minimum value. For low values of current in the pole windings the flux density in the stator/rotor iron is small and the reluctance of these parts of the flux path is much less than the reluctance of the air-gap between the stator and rotor teeth. As the winding current is increased, however, the flux density in the iron eventually reaches its saturation level. Further increases in winding current then produce a diminishing return in terms of improved flux level.

Another limitation on pole field strength arises from the heating effect of the winding currents. The power dissipated in the windings is proportional to the square of the current, so the temperature rise of the windings increases rapidly for higher currents. In most applications it is the ability of the winding insulation to withstand a given temperature rise which limits the current to what is termed its 'rated' value. For a well-designed variable-reluctance stepping motor the limitations on pole flux density and winding temperature rise are both effective (Harris *et al*, 1977); the stator/rotor iron reaches magnetic saturation at the rated winding current.

For the three-stack motor illustrated in Fig. 1.3 there are four poles, and hence four pole windings, per stack. Since all four windings in one stack must be excited concurrently it is common practice to interconnect the windings to form one phase. The three alternative methods of connecting four windings are shown in Fig. 1.4.

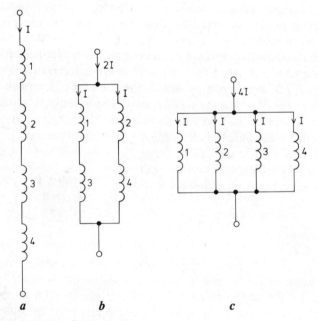

Fig. 1.4   *Interconnection of pole windings*

a   series
b   series/parallel
c   parallel

Although the rated pole winding current depends only on the acceptable temperature rise, the corresponding rated phase current also depends on the interconnection, as shown in Table 1.1:

Table 1.1

Connection	Rated current	Resistance	Rated voltage	Power
Series	I	$4r$	$4rI$	$4rI^2$
Series/parallel	$2I$	r	$2rI$	$4rI^2$
Parallel	$4I$	$r/4$	rI	$4rI^2$

I = Rated pole winding current.

r = Pole winding resistance.

The rated phase voltage is the voltage which must be applied at the phase terminals to circulate the rated current in the windings. For the series connection the phase current is low and the voltage high compared to the parallel connection, but there is no difference in the power supplied to the phase. Most manufacturers produce a given design of stepping motor with a range of winding interconnections, so the user can elect to use a low-voltage, high-current drive with the parallel connection or a high-voltage, low-current drive with the series connection.

1.3   Single-stack variable-reluctance stepping motors

As its name implies, this motor is constructed as a single unit and therefore its cross-section parallel to the shaft is similar to one stack of the motor illustrated in Figs.1.2 and 1.3. However the cross-section perpendicular to the shaft, shown in Fig.1.5, reveals the essential differences between the single- and multi-stack types.

Considering the stator arrangement we see that the stator teeth extend from the stator/rotor air-gap to the back-iron. Each tooth has a separate winding which produces a radial magnetic field when excited by a d.c. current. The motor of Fig.1.5 has six stator teeth and the windings on opposite teeth are connected together to form one phase. There are therefore three phases in this machine, the minimum number required to produce rotation in either direction. Windings on opposite stator teeth are in opposing senses, so that the radial magnetic field in one tooth is directed towards the air-gap whereas in the other tooth the field is directed away from the air-gap. For one phase excited the main flux path lies from one stator tooth, across the air-gap into a rotor tooth, directly across the rotor to another rotor tooth/air-gap/stator tooth combination and returns via the back-iron. As depicted in Fig.1.5, however, it is possible for a small

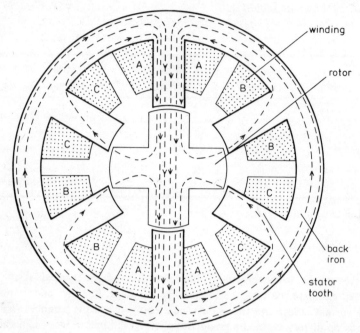

Fig. 1.5 *Cross-section of a single-stack variable-reluctance stepping motor perpendicular to the shaft*
- - - - flux paths for phase *A* excited

proportion of the flux to 'leak' via unexcited stator teeth. These secondary flux paths produce mutual coupling between the phase windings of the single-stack stepping motor.

The most striking feature of the rotor is that it has a different number of teeth to the stator; the example of Fig.1.5 has four rotor teeth. With one phase excited only two of the rotor teeth carry the main flux, but note that the other pair of rotor teeth lie adjacent to the unexcited stator teeth. If the phase excitation is changed it is this other pair of rotor teeth which align with the newly-excited stator teeth. Fig.1.5 shows the rotor position with phase *A* excited, the rotor having adopted a position which minimises the main flux path reluctance. If the excitation is now transferred to phase *B* the rotor takes a step in the anticlockwise direction and the opposite pair of rotor teeth are aligned with the phase *B* stator teeth. Excitation of phase *C* produces another anticlockwise step, so for continuous anticlockwise rotation the excitation sequence is: *A, B, C, A, B, C, A*, ... Similarly clockwise rotation can be produced using the excitation sequence: *A, C, B, A, C, B, A*, ... It is interesting to find that, in the motor illustrated, the rotor movement is in the opposite direction to the stepped rotation of the stator magnetic field!

The step length can be simply expressed in terms of the numbers of phases and rotor teeth. For an *N*-phase motor excitation of each phase in sequence produces *N* steps of rotor motion and at the end of these *N* steps excitation returns to the original set of stator teeth. The rotor teeth are once again aligned with these stator teeth, except that the rotor has moved a rotor tooth pitch. For a machine with p rotor teeth the tooth pitch is $360/p$ degrees corresponding to a movement of N steps, so:

$$\text{Step length} = 360/Np \text{ degrees} \qquad (1.2)$$

In the example of Fig.1.5 there are three phases and four rotor teeth, giving a step length of 30 degrees.

The number of stator teeth is restricted by the numbers of phases and rotor teeth. Each phase is distributed over several stator teeth and, since there must be as many stator teeth directing flux towards the rotor as away from it, the number of stator teeth has to be an even multiple of the number of phases, e.g. in a three-phase motor there can be 6, 12, 18, 24, ... stator teeth. In addition, for satisfactory stepping action, the number of stator teeth must be near (but not equal) to the number of rotor teeth, e.g. a three-phase, 15 degree step length motor is constructed with eight rotor teeth and usually has twelve stator teeth, although the alternative of six stator teeth would produce the same step length.

1.4   Hybrid stepping motors

A hybrid stepping motor has a permanent magnet mounted on the rotor. The main flux path for the magnet flux, shown in Fig.1.6(*a*), lies from the magnet N-pole, into a soft-iron end-cap, radially through the end-cap, across the air-gap, through

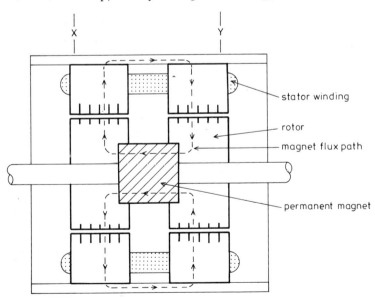

Fig. 1.6   (*a*)   *Cross-section of a hybrid motor parallel to the shaft*

the stator poles of section X, axially along the stator back-iron, through the stator poles of section Y, across the air-gap and back to the magnet S-pole via the end-cap.

There are typically eight stator poles, as in Fig.1.6(b), and each pole has between

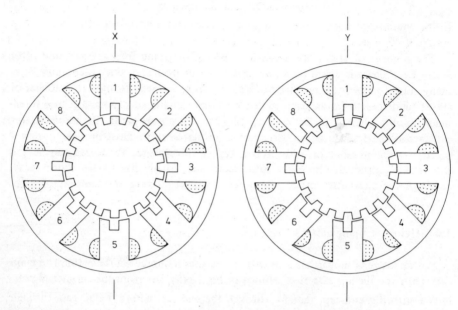

Fig. 1.6   (b)   *Cross-section of a hybrid motor perpendicular to the shaft*

two and four teeth. The stator poles are also provided with windings which are used to encourage or discourage the flow of magnet flux through certain poles according to the rotor position required. Two windings are provided and each winding (phase) is situated on four of the eight stator poles; winding A is placed on poles 1, 3, 5, 7 and winding B is on poles 2, 4, 6, 8. Successive poles of each phase are wound in the opposite sense, e.g. if winding A is excited by positive current the resultant magnetic field is directed radially outward in poles 3 and 7, but radially inward in poles 1 and 5. A similar scheme is used for phase B and the situation for the whole machine is summarised in the Table 1.2.

Table 1.2

Winding	Current direction	Direction of pole magnetic field	
		Radially outward	Radially inward
A	Positive	3, 7	1, 5
A	Negative	1, 5	3, 7
B	Positive	4, 8	2, 6
B	Negative	2, 6	4, 8

The influence of winding excitation on the magnet flux path can be understood by considering the example of winding A excited by positive current. The magnet flux in section X has to flow radially outwards and the excitation of A therefore results in most of the magnet flux flowing in poles 3 and 7. However in Section Y the situation is reversed, since the magnet flux must flow radially inwards and so is concentrated in poles 1 and 5.

Both the stator poles and rotor end-caps are toothed. For the motor illustrated in Fig.1.6 each of the eight poles has two teeth, giving a total of sixteen stator teeth, and the rotor has eighteen teeth. Note that the stator teeth in sections X and Y are fully aligned, whereas the rotor teeth are completely mis-aligned between the two sections. If the magnet flux is concentrated in certain poles because of the winding excitation then the rotor tends to align itself so that the air-gap reluctance of the flux path is minimised. In the example of positive excitation of winding A the stator and rotor teeth are aligned under poles 3, 7 of section X and poles 1, 5 of section Y, as illustrated in Fig.1.6.

Continuous rotation of the motor is produced by sequential excitation of the phase windings. If the excitation of A is removed and B is excited with positive current then alignment of the stator and rotor teeth has to occur under poles 4, 8 of section X and poles 2, 6 of section Y. The rotor moves one step clockwise to attain the correct position. Clockwise rotation can be continued by exciting phase A then phase B with negative current. This sequence can be represented by: $A+, B+,$ $A-, B-, A+, B+, \ldots$. Alternatively anticlockwise rotation would result from the excitation sequence: $A+, B-, A-, B+, A+, B-, \ldots$.

The length of each step can be simply related to the number of rotor teeth, p. A complete cycle of excitation for the hybrid motor consists of four states and produces four steps of rotor movement. The excitation state is the same before and after these four steps, so the alignment of stator/rotor teeth must occur under the same stator poles. Therefore four steps correspond to a rotor movement of one tooth pitch ($= 360/p$ degrees) and for the hybrid motor:

$$\text{Step length} = 90/p \text{ degrees} \qquad (1.3)$$

The motor illustrated in Fig.1.6 has eighteen rotor teeth and a step length of 5 degrees. Hybrid motors are usually produced with somewhat smaller step lengths than this; the motor shown in Fig.1.7 has fifty rotor teeth and a step length of 1·8 degrees.

1.5   Comparison of motor types

The system designer is faced with a choice between hybrid and variable-reluctance stepping motors and his decision is inevitably influenced by the application; it is not possible to state categorically that one type is 'better' in all situations. Hybrid motors have a small step length (typically 1·8 degrees), which can be a great advantage when high resolution angular positioning is required. A survey of manufacturers' data by Harris *et al*. (1977) reveals that the torque producing capability for a given motor volume is greater in the hybrid than in the variable-reluctance

Fig. 1.7   *A commercial hybrid stepping motor (this motor has three basic hybrid motor units, of the type depicted in Fig. 1.6, mounted on a common shaft)*
GEC 156 – 370
[photograph by GEC Small Machines Ltd., Warley, UK]

motor, so the hybrid motor is a natural choice for applications requiring a small step length and high torque in a restricted working space. When the windings of the hybrid motor are unexcited the magnet flux produces a small 'detent torque', which retains the rotor at the step position. Although the detent torque is less than the motor torque with one or more windings fully excited, it can be a useful feature in applications where the rotor position must be preserved during a power failure.

Variable-reluctance motors have two important advantages when the load must be moved a considerable distance, e.g. several revolutions of the motor. Firstly, typical step lengths (15 degrees) are longer than in the hybrid type so less steps are required to move a given distance. A reduction in the number of steps implies less excitation changes and, as we shall see in Chapters 5 and 6, it is the speed with which excitation changes can take place which ultimately limits the time taken to move the required distance. A further advantage, highlighted by Bakhuizen (1976), is that the variable-reluctance stepping motor has a lower rotor mechanical inertia than the hybrid type, because there is no permanent-magnet on its rotor. In many cases the rotor inertia contributes a significant proportion of the total inertial load on the motor and a reduction in this inertia permits faster acceleration (Chapter 6).

Apart from the two basic types of stepping motor discussed in this Chapter, there are available several other devices capable of stepping action. The permanent-magnet stepping motor has a similar stator construction to the single-stack variable-reluctance type, but the rotor is not toothed and is composed of permanent-magnet material. In the example of Fig.1.8 the rotor has two magnetic poles which

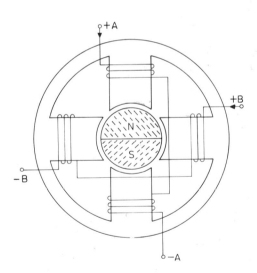

Fig. 1.8 *Permanent-magnet stepping motor*

align with two of the stator teeth according to the winding excitation. A change in excitation between the two windings produces a step of 90 degrees. The current polarity is important in the permanent-magnet motor; the rotor position illustrated is for positive current in winding A, a switch to positive current in winding B would produce a clockwise step, whereas negative excitation of B would give anticlockwise rotation. It is difficult to manufacture a small permanent-magnet rotor with a large number of poles and consequently stepping motors of this type are restricted to step lengths in the range 30–90 degrees. The torque per unit volume of the permanent-magnet stepping motor is relatively poor, so production is limited to the smallest sizes.

At the opposite end of the size range is the electrohydraulic stepping motor, which is used in situations requiring very high torque. The motor is basically a closed-loop hydraulic control system which derives its input from a small conventional electrical stepping motor. Torque gains of several hundred are possible and the electrohydraulic stepping motor has many machine tool applications. It is described in detail by Bell *et al.* (1970) and Kuo (1974).

Chapter 2

Drive circuits

2.1 Introduction

The control signals for a stepping motor system are invariably low power, e.g. TTL digital integrated circuits provide 5 V at 18 mA, whereas a typical variable-reluctance stepping motor giving a torque of 1·2 Nm has a rated winding excitation of 5 V and 3 A. Therefore, if the drive circuit is based on conventional bipolar junction transistor switches, the controller must be interfaced to the motor via several stages of switching amplification. In future the interfacing may be accomplished directly with VMOS power field-effect transistors (Davies, 1980). Many manufacturers provide drive circuits compatible with their motors, but this is one area where the stepping motor user has considerable scope for innovation. A bewildering number of configurations are now available (Sigma Instruments, 1972; Cassat, 1977), but in this Chapter discussion is confined to the basic drive circuits and the potential benefits of more sophisticated drives are examined at a later stage (Chapter 5).

The variable-reluctance stepping motor has at least three phases, but the phase currents need only be switched on or off; the current polarity is irrelevant to torque production. A simple unipolar drive circuit — so-called because it produces unidirectional currents — suitable for use with a variable-reluctance type motor is discussed in Section 2.2. For the hybrid motor, or any type of motor incorporating a permanent-magnet, there are only two phases, but the current polarity is important and a bipolar drive is required to give bidirectional phase currents. The transistor bridge drive introduced in Section 2.3 uses more semiconductor devices per phase than the unipolar drive, but by placing additional windings in the hybrid motor it is possible to simplify the drive; this technique is described in Section 2.4.

2.2 Unipolar drive circuit

A simple unipolar drive circuit suitable for use with a three-phase variable-reluctance stepping motor is shown in Fig.2.1. Each phase winding is excited by a separate drive circuit, which is controlled by a low-power 'phase control signal'.

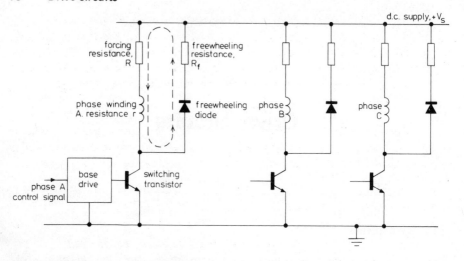

Fig. 2.1   *Three-phase unipolar drive circuit*
- - - - freewheeling current path

This control signal may require several stages of switched amplification before it attains the power level required at the base of the phase transistor.

The phase winding is excited whenever its switching transistor is saturated by a sufficiently high base current. Under these conditions the full d.c. supply voltage is applied across the series combination of phase winding and forcing resistance, since the voltage drop across the saturated transistor is small (typically 0·1 V). The d.c. supply voltage (V_S) is chosen so that it produces the rated winding current (I) when applied to the total phase circuit resistance, which is equal to the sum of the phase winding (r) and forcing (R) resistances:

$$V_S = I\,(r + R) \tag{2.1}$$

In general the phase winding has a considerable inductance, so its natural electrical time constant (inductance/resistance) is long. The build-up of phase current to its rated value would be too slow for satisfactory operation of the motor at high speeds. By adding the forcing resistance, with a proportional increase in supply voltage, the phase electrical time constant can be reduced, enabling operation over a wider speed range. The function of the forcing resistance is given more detailed consideration in Chapter 5.

Another consequence of the finite phase winding inductance is that the phase current cannot be switched off instantaneously. If the base drive of the switching transistor was suddenly removed a large induced voltage would appear between the transistor collector and emitter, causing permanent damage to the drive circuit. This possibility is avoided by providing an alternative current path — known as the freewheeling circuit — for the phase current. When the switching transistor is turned off the phase current can continue to flow through the path provided by the freewheeling diode and freewheeling resistance. If the phase current is established at its rated value then the maximum voltage ($V_{ce\ max}$) across the switching transistor occurs in the instant after the transistor switch is opened. The current (I) has not started to decay and flows through the free-wheeling resistance (R_f), so the maximum collector-emitter voltage (neglecting the forward voltage drop across the freewheeling diode) is:

$$V_{ce\ max} = V_s + R_f I \qquad (2.2)$$

The phase current therefore decays in the freewheeling circuit and the magnetic energy stored in the phase inductance at turn-off is dissipated in the freewheeling circuit resistance (winding + forcing + freewheeling) resistances.

Design example

A three-phase variable-reluctance stepping motor has a total phase winding resistance of 1 ohm and an average phase inductance of 40 mH. The rated phase current is 2 A. Design a simple unipolar drive circuit to give electrical time constants of 2 ms at phase turn-on and 1 ms at turn-off.

$$\text{Electrical time constant} = \text{inductance/resistance} \qquad (2.3)$$

For the turn-on time constant of 2 ms the total phase resistance: = 40/2 ohms = 20 ohms.

Since the winding resistance contributes 1 ohm to the total phase resistance:

Forcing resistance = 19 ohms

This resistance must be able to dissipate the power losses when the phase is continuously excited by the rated current (= 2 A). So:

Power rating = $(\text{current})^2 \times (\text{resistance}) = 76$ W

The d.c. supply voltage can now be found using eqn. (2.1):

Supply voltage = rated current x phase resistance = 2 x 20 V = 40 V

Turning now to the freewheeling circuit, the required time constant for current decay is 1 ms, so the total freewheeling circuit resistance [from eqn. (2.3)] is 40 ohms. Since the total phase resistance is 20 ohms:

Freewheeling resistance = 20 ohms

The power rating of this freewheeling resistance depends on the operating speed of the motor. The energy stored in the phase inductance at turn-off is:

$$\text{Stored energy} = \text{inductance} \times (\text{current})^2/2$$

$$(2.4)$$

$$= 40 \times 4/2 \text{ mJ} = 0.08 \text{ J}$$

and this stored energy is dissipated in the phase and freewheeling resistances. In this case the resistances are equal, so half the stored energy (0.04 J) is dissipated in the freewheeling resistance each time the corresponding phase is turned off. At a speed of 600 steps per second, for example, each of the three phases turns off 200 times per second and the average power dissipated in the freewheeling resistance at this speed is 200 × 0.04 W = 8 W. The analysis of power losses at higher speeds becomes more involved, because the phase current at turn-off is itself a function of operating speed. By assuming that the phase current has reached its rated value, a 'worst case' estimate of the freewheeling resistance power rating is obtained.

Finally the two semiconductor devices can be considered. The freewheeling diode has to withstand a reverse voltage equal to the d.c. supply voltage when the transistor is conducting and it must conduct a peak forward current equal to the rated phase current when the transistor first turns off. The switching transistor must tolerate a collector-emitter voltage given by Eqn. (2.2):

$$V_{ce\ max} = 40 + (20 \times 2) \text{ V} = 80 \text{ V}$$

It has to conduct the phase current (2 A) and must have the highest possible current gain under saturated conditions.

Base I requirement?

2.3   Bipolar drive circuit

One phase of a transistor bridge bipolar drive circuit, suitable for use with a hybrid or permanent-magnet stepping motor, is shown in Fig.2.2. The transistors are

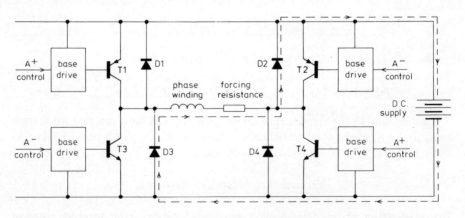

Fig. 2.2   *One phase of a transistor bridge bipolar drive circuit*
     - - - - freewheeling current path after T1 and T4 turn off

switched in pairs according to the current polarity required. For positive excitation of the phase winding transistors T1 and T4 are turned on, so that the current path is from the supply, through transistor T1 to the phase winding and forcing resistance, then through transistor T4 back to the supply. In the opposite case the transistors T2 and T3 are turned on so that the current direction in the phase winding is reversed.

The four switching transistors in the bridge require separate base drives to amplify the two (positive and negative) phase control signals. In the case of the 'upper' transistors (T1 and T2) the base drive must be referred to the positive supply rail, which may be at a variable potential. For this reason the phase control signals to these upper base drives are often transmitted via a stage of optical isolation.

A bridge of four diodes, connected in reverse parallel with the switching transistors, provides the path for freewheeling currents. In the illustration of Fig.2.2 the freewheeling current path, via diodes D2 and D3, corresponds to the situation immediately after turn-off of transistors T1 and T4. The freewheeling path includes the d.c. supply and therefore some of the energy stored in the phase winding inductance at turn-off is returned to the supply. The consequent improvement in overall system efficiency represents a significant advantage of the bipolar bridge drive over the unipolar drive and for this reason most large (> 1 kW) stepping motors, including variable-reluctance types, are operated from bipolar drives.

Freewheeling currents in the bipolar drive decay more rapidly than in the unipolar drive, because they are opposed by the d.c. supply voltage. Therefore it is not necessary to include additional freewheeling resistance in the bipolar bridge drive.

Example

A motor with an average phase inductance of 40 mH and rated phase current of 2A is operated from a bipolar drive with a supply voltage of 40 V and a total phase resistance of 20 ohms. At turn-off estimate the time taken for the phase current to fall to zero and the proportion of the stored inductive energy returned to the supply.

At turn-off the phase current decays exponentially from its initial value of +2 A towards a final value of −2 A. If the phase electrical time constant is T (= 2·0 ms in this case) and turn-off begins at time $t = 0$, the instantaneous current is:

$$i = 2 \cdot 0 \exp(-t/T) - 2 \cdot 0[1 - \exp(-t/T)]$$

$$= -2 \cdot 0 + 4 \cdot 0 \exp(-t/T)$$

which, taking the first two terms of the exponential series expansion, can be approximated by:

$$i = -2 \cdot 0 + 4 \cdot 0(1 - t/T)$$

$$= 2 \cdot 0 - 4 \cdot 0(t/T)$$

If the current falls to zero in time t:

$$0 = 2 \cdot 0 - 4 \cdot 0(t'/T)$$

$$t' = T/2 \cdot 0 = 1 \text{ ms}$$

The instantaneous power returning to the supply is $V \times i$ where V is the supply voltage.

Energy returned to the supply

$$= \int_0^{t'} V \times i \, dt$$

$$= \int_0^{t'} 40 \cdot 0 \times (2 \cdot 0 - 4 \cdot 0 t/T) \, dt$$

$$= (80 \cdot 0 t' - 80 \cdot 0 t'^2/T) = 40 \text{ mJ}$$

Initial energy stored in the inductance $= LI^2/2 = 80$ mJ, so 50% of the stored inductive energy is returned to the supply at turn-off.

2.4   Bifilar windings

The transistor bridge bipolar drive circuit requires four transistor/diode pairs per phase, whereas the simple unipolar drive requires only one pair per phase, so drive costs for a hybrid stepping motor are potentially higher than for the variable-reluctance type; a two-phase hybrid motor drive has eight transistors and diodes, but a three-phase variable-reluctance motor drive has only three transistors and diodes. The bridge configuration has the additional complication of base drive isolation for the pair of switching transistors connected to the positive supply rail. From the viewpoint of drive costs the conventional hybrid motor has a severe disadvantage and therefore many manufacturers have introduced 'bifilar-wound' hybrid motors, which can be operated with a unipolar drive.

A bidirectional current flowing in the hybrid motor windings produces a bi-directional field in the stator poles. With a bifilar winding the same result is achieved by two pole windings in opposite senses, as illustrated for one pole in

Fig.2.3. Depending on the field direction, one of the windings is excited by a unidirectional current; in Fig.2.3 the field produced by a positive current in the conventional arrangement is available by exciting the bilfilar +winding with positive current. The effect of negative current in the conventional winding is then achieved by positive excitation of the bifilar —winding.

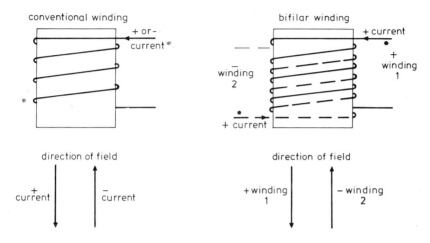

Fig. 2.3   *Comparison of conventional and bifilar windings*

Each of the bifilar pole windings must have as many turns as the original winding and the same rated current, so a bifilar winding has twice the volume of a conventional winding. This additional volume does, of course, increase the manufacturing

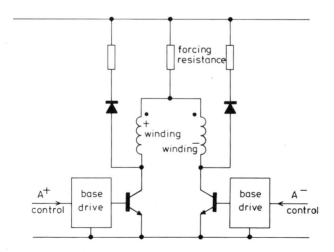

Fig. 2.4   *Unipolar drive circuit for one phase of a bifilar-wound motor*

costs but for small sizes of hybrid motor this is outweighed by the resultant reduction in drive costs.

The two bifilar windings of one phase may be excited by separate unipolar drive circuits of the type discussed in Section 2.2, but one alternative is to 'share' the forcing resistance between the two bifilar windings, as shown in Fig.2.4. There are now only two transistor/diode pairs per phase, so the two-phase hybrid motor with bifilar windings requires four transistors and diodes in its complete drive circuit and has comparable drive costs to a three-phase variable-reluctance motor. The freewheeling path of the bifilar drive does not return energy stored in the inductance at turn-off to the d.c. supply, so the drive has a lower efficiency than the bipolar bridge drive. This reduction in efficiency, coupled with the extra winding costs, is very significant for larger sizes of stepping motor, which are therefore rarely bifilar-wound.

As the two bifilar windings of each phase are situated on the same stator poles within the motor, there is close mutual coupling between the windings and this must be taken into account when considering circuit conditions at turn-on and turn-off. If each winding has N turns then, in the absence of magnetic saturation, the pole flux is proportional to the difference in winding currents:

$$\phi = k_f N(i_1 - i_2) \tag{2.5}$$

The winding flux linkages are then:

$$\lambda_1 = N\phi = k_f N^2 (i_1 - i_2)$$

$$\lambda_2 = -N\phi = k_f N^2 (i_2 - i_1)$$

and for changes in the winding currents i_1 and i_2, the voltages induced in the windings are:

$$v'_1 = d\lambda_1/dt = k_f N^2 (di_1/dt - di_2/dt)$$

$$= L \, di_1/dt - M \, di_2/dt$$

$$v'_2 = d\lambda_2/dt = k_f N^2 (di_2/dt - di_1/dt) \tag{2.6}$$

$$= L \, di_2/dt - M \, di_1/dt$$

where L and M are the winding self and mutual inductances, which have equal magnitude $k_f N^2$.

Because of the mutual coupling between windings, the transient conditions in both of the phase bifilar windings must be considered when calculating turn-on and turn-off times. In the case of a 'shared' forcing resistance, as in the drive circuit of Fig. 2.4, the situation may be further complicated by resistive coupling between the winding circuits.

Example

A bifilar-wound hybrid stepping motor has self and mutual winding inductances of 10 mH and a winding resistance of 1 ohm. Each winding has a separate series forcing resistance of 9 ohms and, with a 30 V drive power supply, the winding current is limited to its rated value of 3 A. Find the winding current time variation at switch-on if the opposite phase winding carries a current of (a) zero and (b) rated, but winding excitation removed.

For separate forcing resistances and equal self and mutual inductances (L) the winding voltage equations are:

$$v_1 = Ri_1 + L\, di_1/dt - L\, di_2/dt \qquad\qquad [2.7(a)]$$

$$v_2 = Ri_2 + L\, di_2/dt - L\, di_1/dt \qquad\qquad [2.7(b)]$$

where R is the total circuit resistance (9 + 1 ohms = 10 ohms).

(a)   Winding 2 is switched-off ($v_2 = 0$) and Eqn. [2.7(b)] reduces to:

$$-Ri_2 = L\, di_2/dt - L\, di_1/dt$$

and substituting into Eqn. [2.7(a)], with $v_1 = V$ (= 30 V):

$$V = Ri_1 + Ri_2 \qquad\qquad (2.8)$$

Immediately before switch-on both winding currents are zero and, from Eqn. (2.5), the pole flux is zero. This flux cannot change instantaneously and therefore at the instant after switch-on $i_1 = i_2$, and substituting this condition into Eqn. (2.8) reveals that the currents in both windings jump to half the rated value, $V/2R = 1\cdot5$ A.

Differentiating Eqn. (2.8) with respect to time:

$$0 = R(di_1/dt + di_2/dt)$$

$$di_1/dt = -di_2/dt$$

and substituting into Eqn. [2.7(a)]:

$$V = Ri_1 + 2L\, di_1/dt$$

which has the solution for $t > 0$:

$$i_1 = (V/2R)[2 - \exp(-Rt/2L)] \qquad\qquad (2.9)$$

Eqn. (2.8) can be solved to give the corresponding expression for i_2:

$$i_2 = (V/2R) \exp(-Rt/2L)] \qquad\qquad (2.10)$$

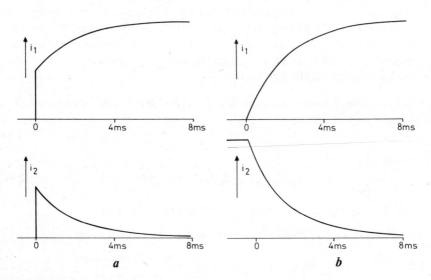

Fig. 2.5 *Bifilar winding currents at switch-on*

    a   opposite winding initially unexcited
    b   opposite winding current initially equal to rated, but winding unexcited

From Eqn. (2.9) and (2.10), we see that after the initial jump to half rated current at switch-on, the winding currents change exponentially with a time constant of $2L/R = 2$ ms. The current waveforms are shown in Fig. 2.5(*a*).

(*b*)   For winding 2 unexcited at switch-on, Eqns. (2.7) become:

$$V = Ri_1 + L\, di_1/dt - L\, di_2/dt \qquad\qquad [2.11(a)]$$

$$0 = Ri_2 + L\, di_2/dt - L\, di_1/dt \qquad\qquad [2.11(b)]$$

Adding these two equations gives:

$$V = Ri_1 + Ri_2 \qquad\qquad (2.12)$$

At switch-on winding 2 carries a current V/R and, by substituting into Eqn. (2.12), we find that $i_1 = 0$ immediately after switch-on, i.e. there is no current jump in this case.

Differentiating Eqn. (2.12) with respect to time:

$$0 = R(di_1/dt + di_2/dt)$$

$$di_1/dt = -di_2/dt$$

and substituting into Eqn. [2.11(a)] :

$$V = Ri_1 + 2L \, di_1/dt$$

which has a solution (for $i_1 = 0$ at $t=0$):

$$i_1 = (V/R)[1 - \exp(-Rt/2L)] \qquad (2.13)$$

Eqn. (2.12) then yields the corresponding expression for i_2 :

$$i_2 = (V/R)\exp(-Rt/2L) \qquad (2.14)$$

So in this case the winding currents undergo a smooth exponential change with time constant $2L/R = 2$ ms, as shown in Fig. 2.5.(b).

Chapter 3

Accurate load positioning:
static torque characteristics

3.1 Introduction

Most stepping motor applications involve accurate positioning of a mechanical load. For example the position of a graph-plotter pen is defined very precisely by the number of switched excitation changes which have taken place in the controlling motor. External load torques, perhaps caused by friction, give rise to a small error in position when the motor is stationary. The motor must develop enough torque to balance the load torque and the rotor is therefore displaced by a small angle from the expected step position. The resultant 'static position error' depends on the external torque, but is independent of the number of steps previously executed; the position error is non-cumulative.

The maximum allowable positional error under static conditions often dictates the choice of motor, so this Chapter deals with the relationship between the static position error and the parameters of the motor, drive and load. In many cases the static error can be reduced if several phases of the stepping motor are excited at the same time, so the potential benefits of multi-phase excitation are examined with reference to both hybrid and variable-reluctance stepping motors. An alternative method of minimising static error is to connect the motor to the load by a gear or, if linear load positioning is required, by a leadscrew, so the effects of these mechanical connections are also investigated.

3.2 Static torque/rotor position characteristics

Manufacturers generally supply information about the torque producing capability of a stepping motor in the form of a graph — known as the static torque/rotor position characteristic — showing the torque developed by the motor as a function of rotor position for several values of winding current. A typical set of curves is shown in Fig. 3.1 for a variable-reluctance motor with one phase excited, although the hybrid motor exhibits similar characteristics.

At the step position the appropriate sets of rotor and stator teeth are completely aligned (see Chapter 1) and no torque is produced by the motor. If the rotor is

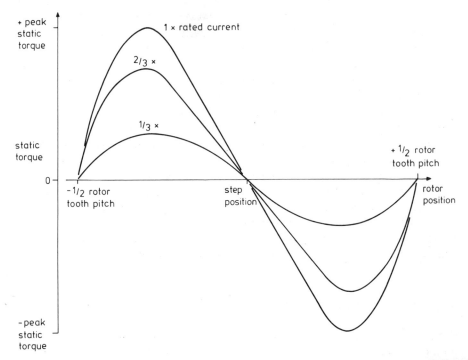

Fig. 3.1   (a)   *Static torque/rotor position characteristics at various phase currents*

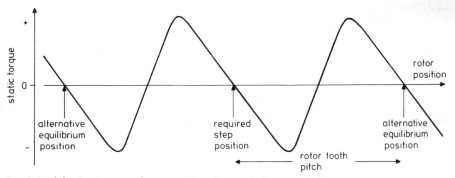

Fig. 3.1   (b)   *Static torque/rotor position characteristic at rated phase current*

slightly displaced from the step position a force is developed between the stator and rotor teeth (Harris *et al.*, 1977) giving a torque which tends to return the rotor to the step position; a rotor displacement in the negative direction produces a positive torque and a positive displacement results in a negative torque. The static torque/rotor position characteristic repeats with a wavelength of one rotor tooth pitch, so the rotor only returns to the correct step position if it is not displaced by more than half a rotor tooth pitch. For larger displacements the rotor and stator teeth become aligned in stable equilibrium at a distance which is a multiple of the rotor tooth pitch from the required step position [see Fig. 3.1(*b*)].

The shape of the static torque/rotor position characteristic depends on the dimensions of the stator and rotor teeth, as well as the operating current. Prediction of the characteristic from the internal geometry of the motor is a complex electromagnetic problem (Ertan *et al.*, 1980) which, from the user's point of view, can be safely left with the motor designer. However it is important to note the relationship between static torque and phase current when the rotor is displaced from the step position. In the absence of magnetic saturation it can be shown (Roters, 1941) that the torque produced is proportional to the square of the phase current in a variable-reluctance motor and linearly proportional to the phase current in a hybrid motor. For most motors the static torque/rotor position characteristic exhibits a rapidly diminishing return in terms of torque produced as the phase current approaches its rated value [see Fig. 3.1(*a*)], indicating that magnetic saturation of the stator and rotor teeth occurs at the higher currents.

The maximum value of static torque is known as the 'peak static torque'. Strictly speaking the peak static torque is a function of phase current, but it is often quoted as a single value corresponding to the rated phase current.

3.3   Position error due to load torque

If an external load torque is applied to the motor then the rotor must adopt a position at which the motor produces sufficient torque to balance the load torque and maintain equilibrium. The maximum torque which the motor can produce, and therefore the maximum load which can be applied under static conditions, is equal to the peak static torque. If the load exceeds the peak static torque the motor cannot hold the load at the position demanded by the phase excitation.

A load torque produces a static position error, which can be deduced directly from the static torque/rotor position characteristic. Fig. 3.2, for example, shows the characteristic for a motor with eight rotor teeth and a peak static torque of 1·2 Nm at rated phase current. With a load torque of 0·75 Nm the motor must move approximately 8 degrees from the step position, until the torque developed balances the load.

An estimate of the static position error can be obtained if the static torque/rotor position characteristic, at the appropriate phase current, is approximated by a sinusoid. For a motor with p rotor teeth and a peak static torque T_{PK} at a rotor displacement θ from the step position, the torque developed by the motor is approximately:

$$T = - T_{PK} \sin p\theta \qquad (3.1)$$

When a load torque T_L is applied the rotor is displaced from the demanded position by the angle θ_e, at which the load and motor torques are equal:

$$T_L = T = - T_{PK} \sin p\theta_e$$

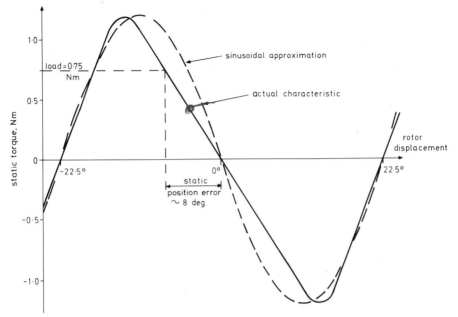

Fig. 3.2   *Derivation of the static position error from the static torque characteristic*

and the static position error is:

$$\theta_e = \frac{\sin^{-1}(-T_L/T_{PK})}{p} \tag{3.2}$$

Therefore the static position error can be reduced by increasing the peak static torque, either by choosing another motor or by using a different excitation scheme, as discussed in the next section. Eqn. (3.2) also shows that a higher number of rotor teeth would be effective in reducing the static position error. Remembering that the step length of a motor is inversely proportional to the number of rotor teeth (Chapter 1), we see that a short step-length motor gives a smaller static position error than an equally-loaded motor with the same peak static torque but longer step length.

Another method of determining the static position error involves the concept of 'stiffness', which is simply the slope of the static torque/rotor position characteristic at the equilibrium position. In Fig. 3.3, for example, the characteristic is approximated by a straight line with slope T', so the motor torque is:

$$T = -T'\theta \tag{3.3}$$

A motor with a high stiffness develops a large torque for a small displacement from equilibrium. For a load torque T_L:

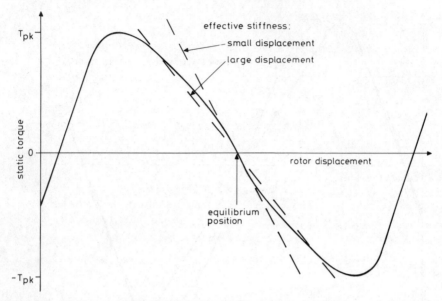

Fig. 3.3   *Derivation of stiffness from the static torque characteristic*

$$T_L = T = - T'\theta$$

and so the static position error is:

$$\theta_e = - T_L/T' \qquad (3.4)$$

In some motors the torque/position characteristic is shaped to give a high value of stiffness near the equilibrium position, so for a given load torque the static position error is reduced. The effective stiffness must then be chosen according to the expected amplitude of the rotor displacement, as shown in Fig. 3.3, in which the stiffness for small loads (up to $0.15T_{PK}$) is appreciably higher than the effective stiffness for larger loads ($0.8T_{PK}$).

3.4   Choice of excitation scheme

The phase windings of both hybrid and variable-reluctance stepping motors are electrically isolated and each phase is excited by a separate drive circuit, so it is possible to excite several phases at any time. This section investigates the potential benefits of multi-phase excitation from the static torque viewpoint. If the peak static torque of a motor can be improved by exciting several phases then the accuracy of load positioning is also improved. Variable-reluctance and hybrid motors are discussed separately in the following sections.

3.4.1   Variable-reluctance motors

The sinusoidal approximations to the static torque/rotor position characteristics shown in Fig. 3.4(*a*) refer to a three-phase multi-stack variable-reluctance stepping motor with one phase excited. There is a mutual displacement between the characteristics corresponding to one step length, so that, for example, the equilibrium position for phase *A* excited is one step length away from the equilibrium position for phase *B* excited. If the rotor is initially at the phase *A* equilibrium position when the excitation is changed to phase *B* the rotor experiences a positive torque which moves the rotor to the phase *B* equilibrium. Conversely if the excitation is changed from *A* to *C* the torque is initially negative, moving the rotor in the negative direction to the phase *C* equilibrium. It is often convenient to think of excitation changes in terms of 'switching off' the static torque/rotor position characteristic of one phase and 'switching on' the characteristic of the next phase.

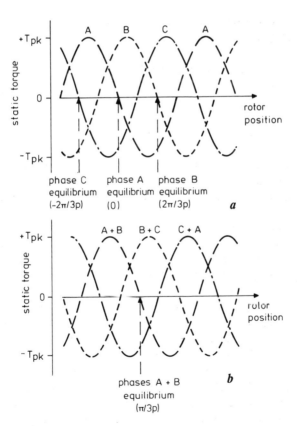

Fig. 3.4   *Static torque/rotor position characteristics for a three-stack variable-reluctance stepping motor*
a   one-phase-on excitation
b   two-phases-on excitation

The effect of exciting two phases at any time is illustrated in Fig. 3.4(*b*), in which the three static torque/rotor position characteristics for phases *AB, BC* and *CA* excited are obtained by summing the torque contributions from each phase at every rotor position. Comparing Figs. 3.4(*a*) and 3.4(*b*) it is apparent that not only are the characteristics still sinusoidal but also the value of peak static torque is unchanged. There is therefore no benefit, as far as static position error is concerned, in exciting a three-phase variable-reluctance motor with two-phases-on, rather than one-phase-on. This result can be confirmed analytically using the following expressions for the three-phase static torque/rotor position characteristics:

$$T_A \quad = -T_{PK}\ \sin(p\theta)$$

$$T_B \quad = -T_{PK}\ \sin(p\theta - 2\pi/3) \qquad\qquad (3.5)$$

$$T_C \quad = -T_{PK}\ \sin(p\theta - 4\pi/3)$$

The resultant torque from two of the phases excited is simply the sum of the separate phase torques:

$$T_{AB} \ = T_A + T_B = -T_{PK}\ \sin(p\theta - \pi/3)$$

$$T_{BC} \ = T_B + T_C = -T_{PK}\ \sin(p\theta - \pi) \qquad\qquad (3.6)$$

$$T_{CA} \ = T_C + T_A = -T_{PK}\ \sin(p\theta - 5\pi/3)$$

Both graphical and analytical results indicate that the only difference between the excitation schemes is in the equilibrium positions; with two phases excited the equilibrium position is between the positions corresponding to separate excitation of each phase.

One further variation is to operate the motor with alternate one- and two-phase-on excitation, i.e. *A, AB, B, BC, C, CA, A,* . . . Each excitation change produces an incremental movement which is half the length of a normal step and therefore this excitation is known as the 'half-stepping' mode of operation.

For simplicity in deriving the analytical results sinusoidal approximations to the characteristics have been used, but nevertheless the general conclusions are valid in most cases. If the torque/position characteristics for one phase excitation is notably non-sinusoidal then the effect of two-phases-on excitation can be checked quite easily using the graphical summation method. For single-stack variable-reluctance stepping motors the direct summation is not strictly valid, since the mutual inductance between phases contributes an additional torque component (Fitzgerald and Kingsley, 1952). Because the phase windings are situated on separate stator teeth in the single-stack motor, the mutual inductance is small and the error introduced in neglecting its torque contribution when several phases are excited is rarely more than 10%.

So far only three-phase variable-reluctance motors have been considered, but for motors with larger numbers of phases then multi-phase excitation is almost obligatory. In the five-phase machine, for example, the highest peak static torque is obtained when two or three phases are excited. This situation is illustrated in Fig. 3.5, where the peak static torque for two or three of the five phases excited is

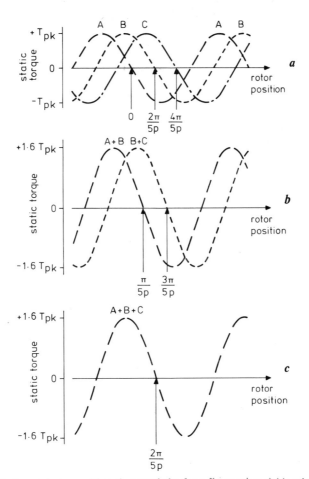

Fig. 3.5 *Static torque/rotor position characteristics for a five-stack variable-reluctance stepping motor*
   a   one-phase-on excitation
   b   two-phases-on excitation
   c   three-phases-on excitation

about 1·6 times the peak static torque with only one phase excited. Once again for two-phases-on there is a shift in the equilibrium positions, which can be confirmed analytically. The static torque/rotor position characteristics for phases *A, B* and *C* of a five-phase motor are represented by:

$$T_A = -T_{PK}\sin(p\theta)$$

$$T_B = -T_{PK} \sin(p\theta - 2\pi/5) \tag{3.7}$$

$$T_C = -T_{PK} \sin(p\theta - 4\pi/5)$$

If two phases (A and B) are excited the total torque is:

$$T_{AB} = T_A + T_B = -2T_{PK} \sin(p\theta - \pi/5) \cos(\pi/5)$$

$$= -1 \cdot 6T_{PK} \sin(p\theta - \pi/5) \tag{3.8}$$

and if three phases (A, B and C) are excited the resultant torque/position characteristic is:

$$T_{ABC} = T_A + T_B + T_C = -1 \cdot 6T_{PK} \sin(p\theta - 2\pi/5) \tag{3.9}$$

Similar results can be obtained for the other combinations of excited phases.

It can be shown generally that the maximum peak static torque is produced in a multi-phase stepping motor when half of the phases are excited (Acarnley *et al.*, 1979), i.e. in a machine with n phases the peak static torque is maximised if:

Number of phases excited $= n/2$ for n even

$$(n+1)/2 \text{ or } (n-1)/2 \quad \text{for } n \text{ odd} \tag{3.10}$$

So a six-phase motor is excited with three-phases-on, while a seven-phase would have either three- or four-phases-on excitation. With odd numbers of phases the motor can be operated in the 'half-stepping' mode, producing the increment of motion equal to half the normal step for each excitation change, e.g. the seven-phase motor is excited alternately three- and four-phases-on, giving the excitation sequence: *ABC, ABCD, BCD, BCDE, CDE, ...*

3.4.2 Hybrid motors

In the hybrid motor there are two phases, which can be excited with positive or negative currents, or, if the motor is bifilar-wound, there are four phases each excited with unipolar current. If each phase is excited in turn four steps are executed while the rotor moves one tooth pitch. Therefore one step length corresponds to a quarter tooth pitch and the four static torque/rotor position characteristics are mutually displaced by this distance, as shown in Fig. 3.6(a). The characteristics are approximated by the sinusoidal functions:

$$T_{A+} = -T_{PK} \sin(p\theta)$$

$$T_{A-} = -T_{PK} \sin(p\theta - \pi)$$

$$T_{B+} = -T_{PK} \sin(p\theta - \pi/2) \tag{3.11}$$

$$T_{B-} = -T_{PK} \sin(p\theta - 3\pi/2)$$

where T_{A+} is the torque produced at rotor position θ when phase A is excited by positive current.

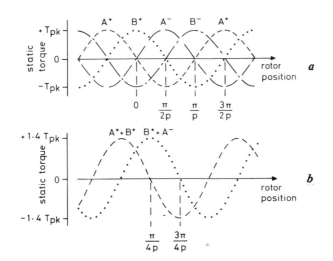

Fig. 3.6   *Static torque/rotor position characteristics for a hybrid motor*
  a   one-phase-on excitation
  b   two-phases-on excitation

The effect of exciting a pair of phases together is shown in Fig. 3.6(*b*) where the peak static torque is improved by a factor of 1·4 over one-phase-on excitation. For phases A and B excited by positive current the total torque is:

$$T_{A+B+} = T_{A+} + T_{B+} = -T_{PK} \sin(p\theta - \pi/4) \cos(\pi/4)$$

$$= -1 \cdot 4 T_{PK} \sin(p\theta - \pi/4) \tag{3.12}$$

and similarly for other phase combinations.

Although the torque produced can often be improved by exciting several phases, it must be remembered that more power is required to excite the extra phases. For example, if two phases of the hybrid motor are excited the power supply must be doubled in capacity, but the torque produced is improved by a factor of only 1·4. This can be an important consideration in applications where the power available to drive the motor is limited.

Imbalance between phases can reduce step accuracy when multi-phase excitation is used and this effect has been investigated by Singh *et al.* (1978).

3.4.3   Mini-step drives

The full-step length of a stepping motor can be divided into smaller increments of rotor motion – known as 'mini-steps' – by partially exciting several phase windings. As an example consider a hybrid stepping motor in which the peak static torque produced by each phase is proportional to the phase current and the phase torques vary sinusoidally with rotor position:

$$T_A = -k_T i_A \sin(p\theta)$$

$$T_B = -k_T i_B \sin(p\theta - \pi/2)$$

(3.13)

For conventional operation the windings are excited by positive and negative rated current, giving a step length of $\pi/2p$ for each excitation change. In this simple case the mini-step drive might produce winding currents which are a fraction of the rated current, I:

$$i_A = I \cos a$$

$$i_B = I \cos(a - \pi/2)$$

(3.14)

The total torque produced by the motor would then be:

$$T = T_A + T_B = -k_T I[\sin p\theta \cos a + \sin(p\theta - \pi/2) \cos(a - \pi/2)]$$

$$= -k_T I \sin(p\theta - a)$$

(3.15)

and the equilibrium position for zero load torque is $\theta = a/p$. Therefore the rotor can be made to take up any intermediate position between the full-step positions ($\theta = 0$, $\pi/2p$, π/p, $3\pi/2p$) if the windings are excited by currents in the correct proportion. A mini-step drive controls the winding current at many levels, so that the rotor moves in 'mini-steps' between the full-step positions, each mini-step corresponding to a change of one level in the winding currents. During one cycle of excitation the current in one phase passes twice through each level and the rotor moves one tooth pitch. Therefore the mini-step length for a drive with N_L current levels is:

$$\text{Mini-step length} = (\text{rotor tooth pitch}/2N_L \qquad (3.16)$$

A hybrid motor with a 1·8 degree full-step length has a rotor tooth pitch of 7·2 degrees (i.e. four full-steps correspond to a rotor movement of one tooth pitch) and a drive with ten current levels would give a mini-step length of 0·36 degrees.

For the conventional full-step drive the equilibrium positions are defined by the alignment of stator and rotor teeth and are therefore independent of current level. However the mini-step positions are critically dependent on the currents in each of the phase windings and any error in current level is translated directly into a position error. The required variation of current levels with rotor position can be deduced from the phase torque/position/current characteristics; for example Eqn. (3.14) shows that the winding currents must vary cosinusoidally with demanded position in the situation where torque is proportional to phase current. Magnetic saturation causes the torque/current and torque/position characteristic to depart from the 'ideal' linear and sinusoidal relationships, but this problem can be counter-acted by adjusting the winding current levels to give uniform mini-steps.

As mini-steps can be made much shorter than full-steps (a typical reduction is twenty mini-steps per full-step) the positional resolution of the stepping motor is improved. However the peak static torque is approximately equal for mini-step and conventional drives, so the positional error due to load torque (see section 3.3) is unaffected and, for large loads, may represent an error of many mini-steps. Stepping motors operated with full-step drives are prone to mechanical resonance (see Section 4.3), but the mini-step drive does much to alleviate this problem by providing a smoother transition between full-step positions (Pritchard, 1976).

The major disadvantage of the mini-step drive is the cost of implementation due to the need for partial excitation of the motor windings at many current levels. Gerber (1975) overcomes this difficulty with a constant supply voltage and several forcing resistances in series with each winding. The resistances can be switched into or out of the circuit according to the phase current required. A more sophisticated drive by Pritchard (1976) incorporates a chopper drive circuit (see Section 5.4.3) in which the reference current level to each phase is changed every mini-step.

3.5 Load connected to the motor by a gear

In some applications a gear is placed between the motor and load with the aim of adjusting the loading on the motor. A schematic diagram of a system incorporating a simple gear train is shown in Fig. 3.7, in which a gear ratio of $N : 1$ is used; N

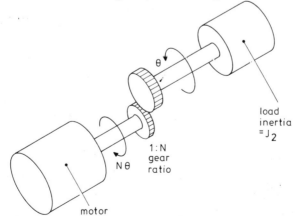

Fig. 3.7   *Motor connected to load by a gear*

revolutions of the motor produce one revolution of the load.

If the load torque is T_L then the torque on the motor is modified to T_L/N, assuming that friction torque effects in the gear are small relative to the load torque. Similarly if the load must be positioned with a maximum static error of θ_{el} then the motor has to operate with a static position error of $\theta_{em} = N\theta_{el}$. As far as static operation is concerned, therefore, there is a considerable advantage in using a high gear ratio to link the motor and load, since the effective load torque at the motor is reduced and the allowable static position error increases compared to the situation where the motor and load are directly connected.

In the example of Fig. 3.2, if the load is connected to the motor by a gear ratio of 3 : 1 then the effective load torque at the motor is 0·75/3=0·25 Nm. From the torque/position characteristic the motor static torque can be deduced; in this case the error is approximately 2 degrees. If the motor moves through a given angle as a result of the load torque then the load itself moves $1/N$ of the angle. In this case the static position error is 2/3 degree, compared to the 8 degree error when the load is directly-coupled, although the effects of friction and backlash in the gear would tend to reduce this improvement.

During dynamic operation if the load is being accelerated then the applied torque is proportional to the angular acceleration and load inertia:

$$T_L = J_L(d^2\theta_L/dt^2) \tag{3.17}$$

At the motor the torque required is:

$$T_m = T_L/N = J_L(d^2\theta_m/dt^2)/N^2 \tag{3.18}$$

since $\theta_m = N\theta_L$. From Eqn. (3.18) the effective inertia at the motor is J/N^2, i.e. the load inertia is reduced by the square of the gear ratio, and it appears that a high gear ratio would enable the motor to accelerate rapidly. This argument is perfectly true, but it is important to remember that one step of motor position produces a load movement which is only a fraction $1/N$ of the motor's step. If the load is to move a fixed distance then the high gear ratio requires the motor to move a large number of steps and for the load movement to be completed in a reasonable time the motor needs to attain a high stepping rate. Coversely with a low gear ratio the effective load inertia is high and the motor accelerates slowly, but has to reach a relatively low stepping rate to move the load at a satisfactory speed.

The above argument can be summarised as follows:

High gear ratio	Low gear ratio
Low reflected inertia	High reflected inertia
Fast acceleration	Slow acceleration
Short load step length	Long load step length
High motor speed	Low motor speed

A high gear-down ratio is an obvious choice where the movement of the load involves substantial periods of acceleration and deceleration. Lower gear ratios are likely to be suitable when the motor's speed capability is restricted.

In some cases the design is complicated by the non-ideal nature of the gear. The present analysis assumes that the gear has negligible inertia and high efficiency, although it is possible to take account of these effects (Tal, 1973). Backlash in the gear can degrade the performance of the system; positional accuracy is reduced and resonance problems (see Chapter 4) are worsened (Ward and Lawrenson, 1977).

3.6   Load connected to the motor by a leadscrew

Some stepping motor applications require incremental linear motion and a number of linear stepping motors, for the direct translation of digital signals into linear steps, have been developed (Finch and Harris, 1979; Langley and Kidd, 1979). However the range of linear stepping motors is restricted and many linear loads are driven from a rotary stepping motor by a leadscrew, which may be an integral part of the motor, as in Fig. 3.8, or a separate component. In the system shown

Fig. 3.8   *A commercial stepping motor with integral leadscrew for use in a floppy-disk drive unit*
[photograph by Warner Electric Inc., USA]

schematically in Fig. 3.9 one revolution of the motor causes a load movement equal to the pitch (h) of the screw, so for an angular movement θ (measured in radians):

$$\frac{\theta}{2\pi} = \frac{x}{h} \tag{3.19}$$

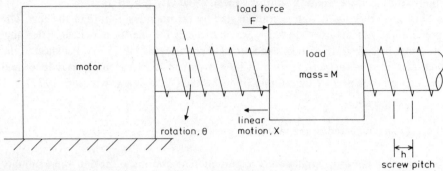

Fig. 3.9   *Motor connected to the load by a leadscrew*

If the load is subject to a force F then, assuming that friction effects in the lead-screw are small, the work done in moving a distance x is Fx and this must equal the work done by the load torque (T_L) at the motor in moving an angle θ:

$$Fx = T_L \theta$$

but x and θ are related by Eqn. (3.19), so:

$$T_L = Fx/\theta = Fh/2\pi \tag{3.20}$$

The static position error for a system subject to a given load force can be calculated as follows:

(a)   calculate the effective load torque at the motor using Eqn. (3.20);
(b)   from the motor static torque/rotor position characteristic calculate the error in the motor's rotational position;
(c)   translate the rotational error into a linear error using Eqn. (3.19).

Example

For a load force of 9 kN and a leadscrew pitch of 0·5 mm calculate the static position error if the stepping motor has the torque/position characteristic shown in Fig. 3.2.

(a)   From Eqn. (3.20)

$$T_L = \frac{9 \times 10^3 \times 0\cdot5 \times 10^{-3}}{2\pi} \text{ Nm} = 0\cdot75 \text{ Nm}$$

(b)   Referring to Fig. 3.2 we see that a load torque of 0·75 Nm produces a rotational error of 8 degrees (= $8 \times 2\pi/360$ radians).

(c)   From Eqn. (3.19)

$$x = \frac{\theta h}{2\pi} = \frac{8 \times 2\pi}{360 \times 2\pi} \times 0.5 \text{ mm} = 0.011 \text{ mm}$$

Static position error for a load force of 9 kN = 0.011 mm

If the load is to be accelerated the force required is proportional to the load mass (M) and the acceleration:

$$F = M(d^2 x/dt^2)$$

This force appears as a load torque at the motor, since from Eqn. (3.20):

$$T_L = Fh/2\pi = Mh(d^2 x/dt^2)/2\pi$$

and substituting for x in terms of θ from Eqn. (3.19):

$$T_L = M\,(h/2\pi)^2\,(d^2\theta/dt^2)$$

Therefore the effective inertia of the load at the motor is:

$$J = M(h/2\pi)^2 \tag{3.21}$$

From the static positioning viewpoint a small screw pitch has the useful property of reducing the load torque at the motor. As far as the dynamic situation is concerned, however, there is a close parallel between the use of a small screw pitch and the high gear-down ratio discussed in the previous section. For a small screw pitch the effective inertia of the load is reduced and the motor can accelerate rapidly, but must attain a high stepping rate to compensate for the small increments of linear movement produced by each motor step.

For simplicity in this analysis the leadscrew has been assumed ideal, i.e. low inertia, high efficiency, low friction and no backlash. All these factors have been taken into account by Tal (1973), who also considered several other motor/load coupling methods, such as a capstan drive and a belt/pulley drive.

Chapter 4

Multi-step operation:
torque/speed characteristics

4.1 Introduction

If a stepping motor is used to change the position of a mechanical load by several steps the system designer needs to know how much torque the motor can produce whilst accelerating, decelerating or running at constant speed. He must ensure that the motor can produce sufficient torque to overcome the load torque and accelerate the load inertia and he will also wish to know the maximum speed at which the motor can drive the load. The necessary information is supplied in the form of a graph – known as the pull-out torque/speed characteristic – showing the maximum torque which the motor can develop at each operating speed. This maximum torque is termed the 'pull-out' torque because if the load torque exceeds this value the rotor is pulled out of synchronism with the magnetic field and the motor stalls. A typical pull-out torque/speed characteristic is shown in Fig. 4.1. In this case the

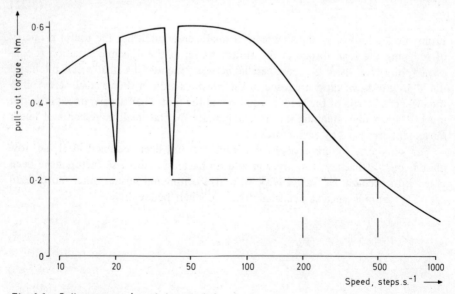

Fig. 4.1 *Pull-out torque/speed characteristic*

motor would be able to drive a load torque of 0·2 Nm at all speeds up to 500 steps per second, because the pull-out torque exceeds the load torque over this speed range. However for a load torque of 0·4 Nm the maximum operating speed would have to be limited to 200 steps per second and there would be additional problems in operating at speeds around 20 and 40 steps per second. For a given load the maximum operating speed is referred to as the 'pull-out' rate, so in this example the pull-out rate for a load of 0·2 Nm is 500 steps per second.

The complete torque/speed characteristic can be divided into several regions, which are discussed separately in this and the next chapter. At 'low' speeds (less than 100 steps per second in the example of Fig. 4.1) the current is quickly established in the windings when a phase is turned on and stays near its rated value for a substantial part of the time for which the phase is excited. The basic pull-out torque/speed characteristic in this region can be deduced from the static torque/rotor position characteristic and the relationship between the two characteristics is discussed in the next section. Two sharp 'dips' in the characteristic of Fig. 4.1 are apparent at speeds near 20 and 40 steps per second. These dips occur in many stepping motor systems and are caused by mechanical resonance in the motor/load combination. Section 4.3 describes this resonance behaviour together with some methods of minimising its effects.

For 'high' speeds (greater than 100 steps per second in Fig. 4.1) the time constant for current rise and decay becomes a significant proportion of the total phase excitation time. The phase current cannot be maintained at its rated value and therefore the torque produced by the motor is reduced. There is a relationship between the motor/drive parameters, operating speed and pull-out torque, but discussion of this topic is reserved for Chapter 5.

4.2   Relationship between pull-out torque and static torque

At a low operating speed the phase current waveforms for a stepping motor are almost rectangular. In Fig. 4.2, for example, the three-phase currents are quickly established at the maximum value because the phase winding time constants (1 ms) are much shorter than the period of each excitation (20 ms at a speed of 50 steps per second). Under these conditions the pull-out torque of the motor can be deduced from the static torque/rotor position characteristics for the particular excitation scheme.

Initially the argument can be simplified by assuming that the motor/load combination has a high inertia, so that variations in motor torque lead to only small changes in motor speed. With this condition of high inertia the pull-out torque is equal to the maximum average torque which can be produced by the motor.

For steady-state operation at zero load torque the position of the rotor at the phase switching times is illustrated in Fig. 4.3 with reference to the torque/position characteristics of a three-phase motor operated with a one-phase-on excitation scheme. Starting at the phase *A* equilibrium position ($\theta=0$) the rated current is established in phase *A* and the rotor has a velocity slightly greater than the

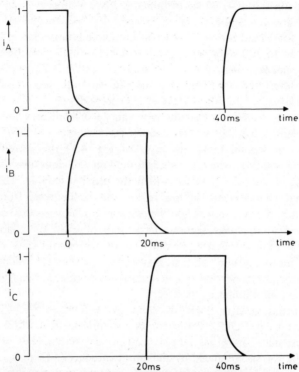

Fig. 4.2   *Current waveforms at 50 steps per second for a three-phase motor operated one-phase-on with a phase winding time constant of 1 ms*

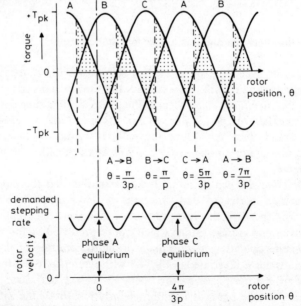

Fig. 4.3   *Rotor position at phase switching times for zero load*

demanded stepping rate. As the rotor moves forward the torque produced by the motor is negative and so the system decelerates until the rotor reaches the position $\theta=\pi/3p$ where the velocity is a minimum. At this point the excitation switches almost instantaneously from phase A to phase B and the motor now produces a positive torque, causing the system to accelerate towards the phase B equilibrium position at $\theta=2\pi/3p$. The motor velocity increases under the influence of the positive torque and the equilibrium position is attained with maximum velocity. The cycle then repeats with the motor producing a negative torque after passing the equilibrium position and the excitation switching from phase B to phase C at the position $\theta=\pi/p$. During the excitation of each phase the motor produces equal positive and negative torques, so there is no net torque production. The motor torque balances the (zero) load torque and therefore there is no resultant torque to accelerate the rotor, which continues at the average speed equal to the demanded stepping rate.

The system is in stable equilibrium because a small increase in the load torque retards the rotor, so the excitation changes occur at slightly smaller rotor positions. During excitation of each phase the motor then produces more positive than negative torque and in the new steady-state the average motor torque again balances the load torque.

Now consider the effect of applying a load torque equal to the pull-out value, so that the motor has to produce the maximum available torque. Fig. 4.4 shows that

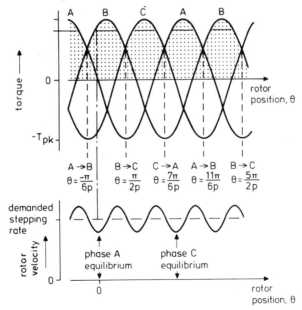

Fig. 4.4   *Rotor position at phase switching times for pull-out load*

the rotor is retarded by the load torque, so that the motor produces a positive torque throughout each excitation period. The equilibrium position for the excited phase is never attained because more torque can be produced by switching to the next phase as the step position is approached. For example excitation is now transferred from phase A to phase B at $\theta = -\pi/6p$. There is a part of each step during which the motor torque is less than the load torque, so that the system decelerates. In the middle of each step the motor torque is near its maximum value, which is greater than the load torque, and the system accelerates, as shown in Fig. 4.4.

With pull-out load torque applied the system is in unstable equilibrium, since any small increase in load retards the rotor, but this now leads to a reduction in motor torque. The difference between the load and motor torques becomes larger and, unless the load torque is reduced quickly, the rotor drops further behind the demanded position until synchronism is lost and the motor stalls.

Having established the rotor positions at which switching must occur to maximise the torque, it is possible to deduce the pull-out torque. In Fig. 4.4 phase A has a torque/position characteristic:

$$T_A = -T_{PK} \sin(p\theta)$$

and phase A must switch on at $\theta = -5\pi/6p$ and switch off at $\theta = -\pi/6p$ for maximum torque. Hence the pull-out torque for one-phase-on operation of the three-phase motor is simply the average of T_A over this interval:

$$\text{Pull-out torque} = \frac{\displaystyle\int_{-5\pi/6p}^{-\pi/6p} -T_{PK} \sin(p\theta)\, d\theta}{(-\pi/6p) - (-5\pi/6p)}$$

$$= 0 \cdot 83\, T_{PK} \tag{4.1}$$

Similar methods can be applied to the calculation of pull-out torque for other excitation schemes.

The dependence of peak static torque on excitation scheme is highlighted in Section 3.4, so a natural question is: 'Which excitation scheme must be chosen to maximise the pull-out torque?'. Referring again to Fig. 4.4 it is apparent that the total torque produced by the motor is maximised if each phase is excited whenever it can contribute a positive torque component. Since the torque/position characteristics are symmetrical about the zero torque axis, each phase must be excited for half of the total excitation sequence. This corresponds to the 'half-stepping' mode in motors with odd numbers of phases, e.g. the three-phase motor is excited in the sequence: A, AB, B, BC, C, CA, A, ... A timing diagram for this phase switching is shown in Fig. 4.5. Motors with even numbers of phases must have half of the phases excited at any time. For a hybrid motor the pull-out torque is maximised with the two-phases-on excitation sequence: $A+B+$, $B+A-$, $A-B-$, $B-A+$, $A+B+$, ...

The maximum pull-out torque can be calculated for an n-phase motor with a

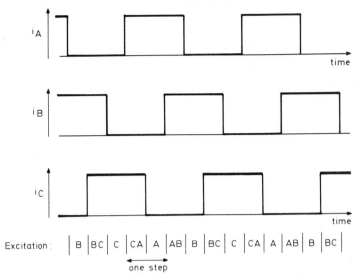

Fig. 4.5   *Half-stepping excitation of a three-phase motor*

peak static torque of T_{PK} when one phase is excited. Each phase is excited for half of the rotor movement in the cycle and therefore contributes an average torque:

$$\text{Phase torque} = \frac{\displaystyle\int_{-\pi/p}^{0} -T_{PK}\sin(p\theta)\,d\theta}{2\pi/p}$$

$$= T_{PK}/\pi$$

and so the total torque from all n phases excited is:

$$\text{Maximum pull-out torque} = nT_{PK}/\pi \qquad (4.2)$$

The assumption that a sinusoid is a good approximation to the shape of the torque/ position characteristic has been made in this derivation. If the actual characteristic shows a significant deviation from a sinusoid an equivalent result can be obtained by evaluating the above integral by a 'counting squares' method.

The pull-out torque/speed characteristic of Fig. 4.1 shows a reduction in pull-out torque from its maximum (0·6 Nm at 60 steps per second) as the operating speed decreases. Aside from the sharp dips, which are discussed in the next section, there is a gradual decline in pull-out torque to 0·48 Nm at a speed of 10 steps per second. The cause of this effect is the oscillation in rotor velocity shown in Figs. 4.3 and 4.4, where it is assumed that the system inertia is sufficient to maintain rotation even if the load torque temporarily exceeds the motor torque. For a system with a relatively low inertia this assumption may not be true and the rotor can stop whenever the load exceeds the motor torque. The effective pull-out torque then corresponds to the 'crossover' point of the static torque/rotor position characteristics, as illustrated in Fig. 4.6. It is at this crossover position that the motor's torque producing capability is at a minimum; at any other position more torque can be produced provided the appropriate phase is excited.

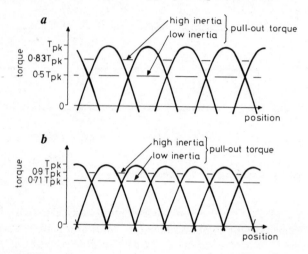

Fig. 4.6   *Pull-out torque for high and low inertia loads in terms of static torque/rotor position characteristics*
a   three-phase motor
b   four-phase motor

For motors with a large number of phases the torque reduction at low speeds is less pronounced. In Fig. 4.6(*a*), for example, with one-phase-on excitation of a three-phase motor the crossover point occurs at a torque of $0.5T_{PK}$, which is 0.6 times the pull-out torque with high inertia ($0.83T_{PK}$). With a four-phase motor excited one-phase-on, however, the pull-out torque for low inertia is $0.71T_{PK}$ and for high inertia is $0.9T_{PK}$, so the pull-out torque is only reduced by a factor of 0.79, as shown in Fig. 4.6(*b*). As the number of phases increases the reduction factor tends towards unity and the system inertia has less effect on the low-speed pull-out torque. The reduction in pull-out torque occurs when the stepping rate is less than the natural frequency of mechanical oscillations for the system, which can be determined from the results of Section 4.3.

4.3   Mechanical resonance

4.3.1   The mechanism of resonance

At very low stepping rates the motor comes to rest at the appropriate equilibrium position after each excitation change. The response of the system to each excitation change — known as the single-step response — is generally very oscillatory; a typical response is shown in Fig. 4.7. In applications requiring frequent accurate positioning

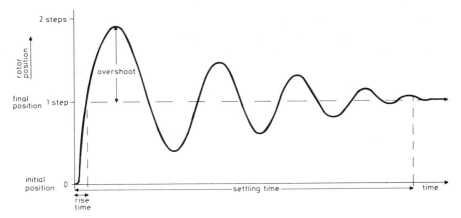

Fig. 4.7 *Typical single-step response*

this poorly-damped response can be a great disadvantage. For example, if a stepping motor is used to drive the carriage of a teletype then the system must come to rest for the printing of each letter. The operating speed of the teletype is limited by the time taken for the system to settle to within the required accuracy at each letter position.

The frequency of oscillation can be predicted for any motor/load combination from the static torque/rotor position characteristic, provided the system is lightly damped (Bakhuizen, 1973). At a rotor position θ from the equilibrium position the motor torque is $-T'\theta$, where T' is the stiffness of the torque/position characteristic. If there is no load torque then this motor torque is used to accelerate the motor/load inertia (J), therefore:

$$-T'\theta = J(d^2\theta/dt^2)$$

$$J(d^2\theta/dt^2) + T'\theta = 0 \qquad (4.3)$$

This is an equation of simple harmonic motion for the rotor position and so the natural frequency of rotor oscillation about the equilibrium position is:

$$\text{Natural frequency,} \quad f_n = (T'/J)^{1/2}/2\pi \qquad (4.4)$$

The simple analysis of oscillation frequency assumes that the system is undamped. In practice there is a small amount of viscous friction present in the system so that the oscillations are lightly damped and the rotor eventually settles at the equilibrium position, as illustrated in Fig. 4.7. Friction effects in an electromechanical system are generally undesirable, since they lead to wear in the moving parts, and are variable, because they are a function of this wear. The designer attempts to reduce friction as far as possible, so most stepping motor systems have very little inherent damping and consequently a poorly-damped single-step response.

The parameters of the single-step response are defined in Fig. 4.7. Rise time is the time taken for the motor to first reach the demanded step position, which is attained with maximum velocity. The system therefore overshoots the target and the amplitude of this first overshoot is expressed as a percentage of the total step, giving the % overshoot. Finally the settling time is the time taken for oscillation to decay so that the system is within 5% of the target.

Once consequence of the highly oscillatory single-step response is the existence of resonance effects at stepping rates up to the natural frequency of rotor oscillation. Fig. 4.8 shows two responses of a motor to a series of steps at different

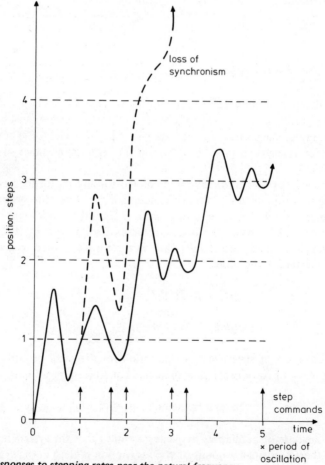

Fig. 4.8   *Responses to stepping rates near the natural frequency*
 — — — — stepping rate = 0·6 x natural frequency
 ———— stepping rate = natural frequency

rates. In the first response the stepping rate is about 0·6 times the natural frequency and therefore the rotor is behind the equilibrium position and has a low velocity when the next excitation change occurs. The rotor quickly settles into a uniform response to each step. In the other response the stepping rate is approximately equal to the natural frequency and so the rotor is at the equilibrium position with a positive velocity at the end of the first step. As a result of this initial velocity the response to the second step is more oscillatory; the rotor swings still further from the equilibrium position. The rotor oscillations increase in amplitude as successive steps are executed until the rotor lags or leads the demanded step position by more than half a rotor tooth pitch. Once this oscillation amplitude is exceeded the motor torque causes the rotor to move towards an alternative step position which is a complete rotor tooth pitch from the expected position [see Fig. 3.1(*b*)]. The correspondence between rotor position and the number of excitation changes is now lost and the subsequent rotor movement is erratic. Note that motors with a large number of phases have an advantage here, since a step length is a small proportion of the rotor tooth pitch [Eqn. (1.1)] and therefore in these motors the rotor can be several steps from the demanded position without losing synchronism.

This resonant behaviour of the system leads to a loss of motor torque at well-defined stepping rates, as illustrated by the dips in the pull-out torque/speed characteristic of Fig. 4.1. The stepping rates at which these dips are likely to occur can be predicted if the natural frequency is known either from Eqn. (4.4) or from direct measurement of the single-step response. Resonance is likely to occur if, at the end of the excitation interval, the rotor is in advance of the equilibrium position and has a positive velocity. These regions are indicated in Fig. 4.9. The

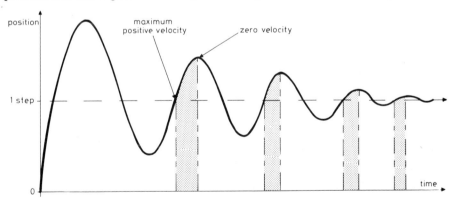

Fig. 4.9 *Regions of the single step response in which phase switching leads to resonance*

rotor has to pass through these regions after times which are a multiple of the rotor oscillation period $(1/f_n)$ and therefore:

$$\text{Resonant stepping rates} = f_n/k = (T'/J)^{1/2}/2\pi k, \quad k = 1, 2, 3, \ldots \quad (4.5)$$

A motor with a natural frequency, from Eqn. (4.4), of 100 Hz can be expected to have dips in the torque/speed characteristics at 100, 50, 33, 25, 20,... steps per second. This result is not precise because the oscillation frequency depends on the amount of damping, but it is sufficiently accurate for most purposes. The additional complication of damping-dependent oscillation frequency is included in the analysis by Lawrenson and Kingham (1977).

For applications requiring repeated fast positioning over a single-step (e.g. the teletype carriage positioning application mentioned at the beginning of this section) it is possible to utilise the high overshoot of the system. If the step corresponds to a change of excitation from phase *A* to phase *B*, for example, the half-step with both phases *A* and *B* excited is first taken. Fig. 4.10 shows that the system overshoots

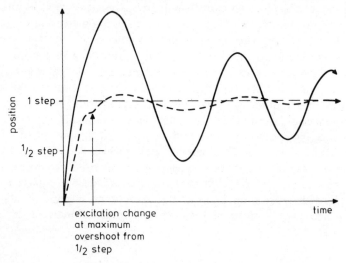

Fig. 4.10   *Intermediate half-step response*
——————— response to full step excitation change
— — —— response with intermediate half-step

the demanded position for *A* and *B* excited, coming to rest near the phase *B* equilibrium position. At this time the excitation is switched to phase *B* only and the transition to the final step is accomplished from a small initial error with consequently small overshoot. The contrast between this response and the effect of changing directly from single-phase excitation of *A* to *B* is shown in Fig. 4.10. Unfortunately the timing of the excitation changes in this intermediate half-step control is quite critical and is heavily-dependent on load conditions. It is therefore restricted in application to situations where the load is constant.

The resonant tendencies of a stepping motor system can be reduced by introducing more damping and therefore limiting the amplitude of oscillation in the single-step response. There are two important techniques for improving the damping, using either mechanical or electrical methods, and these are discussed in the following sections.

4.3.2 The viscously-coupled inertia damper

One mechanical method of damping the single-step response is to introduce additional viscous friction (torque proportional to speed), so that the rotor oscillations decay at a faster rate (Kent, 1973). However the use of straightforward viscous friction is undesirable because the operation of the motor at high speeds is severely limited by the friction torque. A solution to this problem is the viscously-coupled inertia damper (VCID), sometimes known as the Lanchester damper. This device gives a viscous friction torque for rapid speed changes, such as occur in the single-step response, but does not interfere with operation at constant speeds.

The essential features of the VCID are illustrated in Fig. 4.11. Externally the

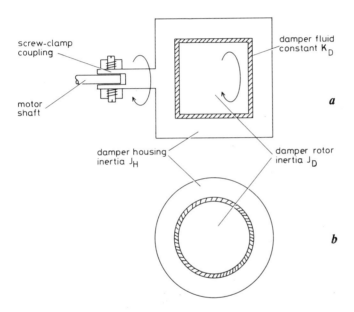

Fig. 4.11 *Cross-sections of the viscously-coupled inertia damper*
a parallel to the shaft
b perpendicular to the shaft

damper appears as a cylindrical inertial load which can be clamped to the motor shaft so the damper housing rotates at the same speed as the motor. Internally the damper has a high inertia rotor which is separated from the housing by a viscous fluid. The housing and the inner rotor can therefore rotate relative to each other, but are loosely coupled by the viscous fluid. When relative motion occurs between the damper components there is a mutual drag torque.

The basic parameters of the damper are its inertia (J_D) and the viscous fluid constant k_D, but in addition there is the inertia of the damper housing (J_H), which increases the overall motor/load inertia. In terms of these parameters the drag torque (T_D) can be expressed in terms of the difference between housing and rotor velocities:

$$T_D = k_D[(d\theta/dt) - (d\theta_D/dt)] \tag{4.6}$$

where θ is the instantaneous position of the motor/load/housing and θ_D is the damper inertia position. Dampers are carefully designed so that this linear relationship is preserved over a wide range of speed difference. The damper torque acts as a drag torque on the motor shaft and also accelerates the damper rotor:

$$T_D = J_D(d^2\theta_D/dt^2) \tag{4.7}$$

If the motor is operating at constant speed the damper rotor must also be running at a constant speed, so from Eqn. (4.7) the damper torque is zero. From Eqn. (4.6) if the damper torque is zero, the damper rotor and housing must be operating at equal speeds.

A well-designed damper can produce a considerable improvement in the single-step response. If the inherent friction torque of the system can be neglected the equation for the rotor position relative to equilibrium (4.3) is modified by the damper torque:

$$T'\theta + J(d^2\theta/dt^2) + T_D = 0 \tag{4.8}$$

where J includes the damper housing inertia, J_H.

Substituting for T_D from Eqn. (4.6):

$$T'\theta + J(d^2\theta/dt^2) + k_D(d\theta/dt) - k_D(d\theta_D/dt) = 0$$

θ_D can be eliminated by differentiating this expression and substituting from Eqn. (4.7):

$$T'(d\theta/dt) + J(d^3\theta/dt^3) + k_D(d^2\theta/dt^2) - k_D T_D/J_D = 0$$

Finally an expression wholly in terms of rotor position can be obtained by substituting for T_D from Eqn. (4.8):

$$T'(d\theta/dt) + J(d^3\theta/dt^3) + k_D(d^2\theta/dt^2) + k_D T'\theta/J_D + k_D J(d^2\theta/dt^2)/J_D = 0$$

and therefore:

$$(d^3\theta/dt^3) + k_D(1/J + 1/J_D)(d^2\theta/dt^2) + (T'/J)(d\theta/dt) + (k_D T'/JJ_D) = 0 \tag{4.9}$$

So the single-step response of a system with a VCID is third-order, compared to the second-order Eqn. (4.3) for the system without a damper.

Turning now to design it is important to answer the question of which damper produces the 'best' single-step response from a given system. In terms of the parameters, we have to choose J_D and k_D for the damper, given J and T' for the motor/load. One problem here is to establish a suitable criterion to judge the quality of a single-step response. Faced with this problem, Lawrenson and Kingham (1974) chose to minimise the integral-with-time-of-absolute-error (ITAE), which corresponds to minimising the area shown in Fig. 4.12. Using this criterion it was

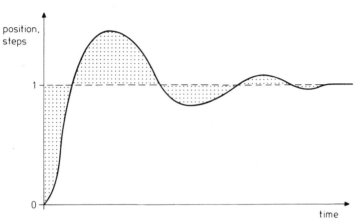

Fig. 4.12 *Single-step response showing the area corresponding to the integral with time of the absolute error (ITAE)*

possible to show that the damper inertia should be four times the total motor/load inertia (including the housing inertia):

$$J_D = 4 \times J \qquad (4.10)$$

and the viscous fluid constant should then be related to the stiffness and damper inertia by:

$$k_D = 0.53(T'J_D)^{1/2} \qquad (4.11)$$

If a well-matched damper is coupled into the system the improvement in the single-step response is achieved with a shorter settling time and lower overshoot, as shown in Fig. 4.13. However the damper does have the effect of increasing the rise time of the response, because the available motor torque has to accelerate both the load and damper inertias towards the step position.

When a VCID is used the total motor/load inertia includes the damper housing inertia and therefore it is important that this inertia be as small as possible. The penalty to be paid for the use of a VCID is that the system is slower to accelerate. Even if the viscous coupling is low (so that the damper housing and rotor operate almost independently) the system inertia is increased by the housing inertia and the acceleration is correspondingly reduced. With a high value of k_D the damper

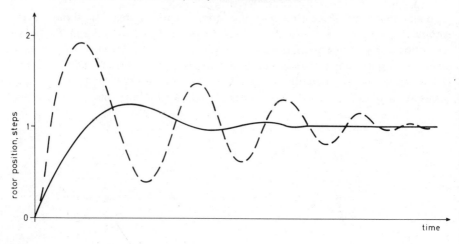

Fig. 4.13   *Effect of VCID on the single-step response*
———————— with VCID
— — — — without VCID

housing and rotor are closely coupled, so the effective system inertia is increased by both the housing and damper rotor inertia.

4.3.3   Electromagnetic damping

The basic aim of any damping scheme is to extract stored mechanical energy, which is in the form of rotational kinetic energy when the system inertia is moving and potential energy when the system is displaced from its equilibrium position. Damping with the VCID is achieved by transferring the system's mechanical energy to and fro between the damper housing and rotor using an inefficient method of coupling (the viscous fluid), so that some of the energy is dissipated with each transfer. Therefore in the VCID the mechanical energy is used to heat the coupling fluid. In electromagnetic damping schemes the mechanical energy acquired by the system in moving between the step positions is transferred to the motor's electrical circuit and dissipated in the motor winding and forcing resistances.

The transfer of energy to the electrical circuit is accomplished by means of the voltages induced in the phase windings when the rotor oscillates, so these voltages are considered first. Fig. 4.14(*a*) shows the variation of magnet flux linked with the two phase windings of a hybrid motor as the rotor position varies over a rotor tooth pitch. This characteristic is approximately sinusoidal with a wavelength equal to the rotor tooth pitch and the sinusoids for the two phases are displaced by $\pi/2p$.

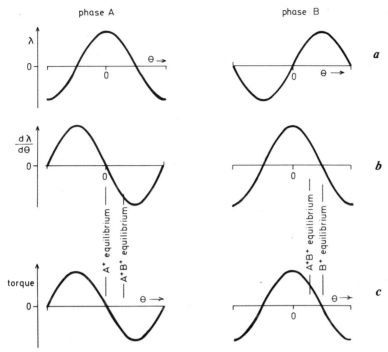

Fig. 4.14   *Flux linkage/rotor position characteristics*
a   flux linkages vs. rotor position
b   rate of change of flux linkages vs. rotor position
c   static torque vs. rotor position

The rate of change of flux linkages with rotor position is shown in Fig. 4.14(*b*), which has the correct phase relationship to Fig. 4.14(*a*). When the flux linkages in phase *A* are at a maximum, for example, the rate of change of flux linked with phase *A* is zero. For a given phase current the torque produced by one phase is proportional to the rate of change of flux linkages with rotor position (Fitzgerald and Kingsley, 1952) and therefore the static torque/rotor position characteristics for positive excitation of the two phases can be deduced, as in Fig. 4.14(*c*).

With only one of the phases excited the rotor moves to the phase equilibrium position, where the torque is zero and, from Fig. 4.14(*b*), the rate of change of flux linkages is also zero. If the rotor oscillates about this one-phase-on equilibrium position the flux linked with the phase winding undergoes only small changes and the voltage induced by the magnet flux is insignificant. Now consider the situation when two phases are excited. The equilibrium position is between the two separate phase equilibrium positions and so the rate of change of flux linkages with rotor position is relatively large. Therefore if the rotor is oscillating about the two-phases-on equilibrium position a voltage is induced in each phase by the magnet flux with a frequency equal to the frequency of rotor oscillation. It is these induced voltages which are used to extract energy from the mechanical system and provide electromagnetic damping.

A rigorous analysis of the mechanisms involved in electromagnetic damping has been undertaken by Hughes and Lawrenson (1975), who demonstrated that the single-step response is third-order when the electrical circuit is taken into account. The results of this analysis show that for a well-damped response the phase resistance (winding + forcing) must be set at an optimum value which depends on several parameters of the motor and load. With two-phases-on excitation damping occurs because the induced voltages produce additional a.c. components of phase current, which are superimposed on the steady d.c. phase current. These a.c. current components give extra power losses in the phase resistance when the rotor is oscillating, so mechanical energy is extracted from the system to supply this extra power. If the phase resistance is set too high the a.c. current component is low and the power losses ($i^2 r$) are small. Conversely if the phase resistance is below the optimum value the a.c. current is high, but there is very little resistance in which the current can dissipate power.

For a hybrid stepping motor the optimum value of phase resistance for maximum electromagnetic damping is:

$$R = (T'/J)^{1/2} \times L \times (1 + k/2) \tag{4.12}$$

where L is the inductance of the phase winding. The factor k is a parameter of the motor and depends on the ratio of the magnet flux linking the phase winding to the flux linkages brought about by the winding current. Typical values of k are in the range 0·25–1·0. A similar result to Eqn. (4.12) also applies to variable-reluctance motors, except that the parameter k has a different definition.

Although the optimum phase resistance can be calculated, in practice it is a fairly simple matter to determine the optimum experimentally. The single-step response can be examined over a range of forcing resistance values (with appropriate changes of supply voltage to maintain constant phase current) until a suitable response is obtained. The discussion has centred on the two-phase hybrid motor, but electromagnetic damping can be produced in all types of motor provided more than one phase is excited when the rotor is settling to the equilibrium position. In some cases the electromagnetic damping effect can be enhanced by introducing a d.c. bias to all phases of the motor (Tal and Konecny, 1980).

As with the VCID, the design of a system for good damping using electromagnetic methods is often in direct conflict with the demands of high-speed operation. In the next Chapter, for example, it is shown that the system requires a large forcing resistance to operate at the highest speeds and in most cases the total phase resistance is then much greater than the optimum for electromagnetic damping. The system designer is therefore left to make a compromise choice of forcing resistance according to the application.

Chapter 5

High-speed operation

5.1 Introduction

In many applications the motor must be able to produce a large pull-out torque over a wide range of stepping rates, so the time taken to position a load is minimised. For example, suppose a motor with the torque/speed characteristic shown in Fig.5.1 has to move a load 1000 steps. If the load torque is 0·5 Nm then

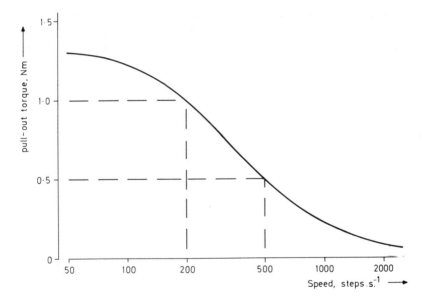

Fig. 5.1   *Maximum operating speeds for different load torques*

the pull-out rate is 500 steps per second and the load is positioned in approximately 1000/500 = 2 seconds. However for a load torque of 1 Nm the maximum speed would have to be restricted to 200 steps per second and the positioning time would be 1000/200 = 5 seconds. Clearly the designer of the system with a load torque of 1 Nm would like to know what parameters of the motor and drive need to be changed so that a pull-out torque of 1 Nm is available at 500 steps per second.

At high stepping rates each phase is excited for only a short time interval and the build-up time of the phase current is a significant proportion of the excitation interval. When a motor is operating at the highest speeds the current in each phase may not even reach its rated value before the excitation interval finishes and the phase is turned off. In addition the time taken for the phase current to decay becomes important at high speeds, because the phase current continues flowing (through the freewheeling diode) beyond the excitation interval dictated by the drive transistor switch. Consequently the pull-out torque falls with increasing stepping rate for two basic reasons:

(*a*) the phase currents are lower, so the motor torque produced at any rotor position is reduced,

(*b*) phase currents may flow at rotor positions which produce a negative phase torque.

A quantitative treatment of these effects for both hybrid and variable-reluctance motors is presented in this chapter.

The calculation of pull-out torque at high speeds is complicated by the variations in current during the excitation time of each phase, which means that there is no longer a simple relationship between the static torque/rotor position characteristic and the pull-out torque. Typical phase current waveforms for one-phase-on unipolar excitation of a three-phase variable-reluctance motor are shown in Fig.5.2. At the lowest operating speeds [Fig.5.2.(*a*)] the current

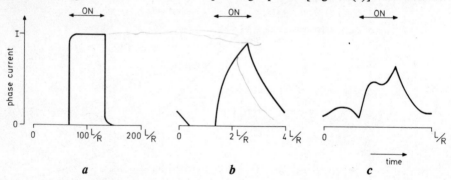

Fig. 5.2 *Typical current waveforms for one-phase-on unipolar excitation of a three-phase motor*
a low-speed
b medium-speed
c high-speed

waveforms are nearly rectangular, the build-up of current to the rated level occupies a minor portion of the excitation time and the methods of Chapter 4 can be used to calculate the pull-out torque. For stepping rates where the phase is only excited for a time similar to the winding time constant, however, the waveform [Fig.5.2(*b*)] is considerably distorted by the nearly exponential rise and decay of the phase current.

At very high operating speeds the voltage induced in the phase windings by the rotor motion must also be considered. The effect of these induced voltages can be seen in the high-speed waveform of Fig.5.2(c), in which the waveform can no longer be described in terms of a simple exponential rise and decay. Even while the phase is switched on it is possible for the current to be reduced by the induced voltage, which is at its maximum positive value when the phase is excited. Similarly when the phase is turned off the decay of current can be temporarily reversed as the induced voltage passes through its maximum negative value. Therefore analysis of the complete pull-out torque/speed characteristics must include the effects of voltages induced in the windings by the moving rotor.

In most stepping motor systems the winding time constant is much less than the period of rotor oscillations about each equilibrium position. At the stepping rates considered in this Chapter we are justified in regarding the rotor velocity as constant; the system inertia is sufficient to maintain a steady speed, even if the motor torque varies slightly during each step. This is commonly known as 'slewing' operation because the rotor is moving continuously without coming to rest at each equilibrium position. It is, of course, possible to take account of the speed variations (Pickup and Tipping, 1976), but this is an unnecessary complication in the evaluation of pull-out torque.

Magnetic saturation has a great influence on the torque produced by a stepping motor (Fig.3.1) and its effects can be included in pull-out torque calculations (Acarnley and Hughes, 1981). However analytical results can only be obtained if saturation is neglected, so this simplification has been used here. Hybrid and variable-reluctance motors receive separate treatment because there are basic differences in the analysis of the two types.

5.2 Pull-out torque/speed characteristics for the hybrid motor

5.2.1 Circuit representation of the motor

The first stage in calculating the pull-out torque is to establish a suitable model for the phase circuits and then use this model to find how the phase current varies with stepping rate.

A hybrid motor has two phase windings, which are mounted on separate stator poles (Chapter 1) and therefore the phase circuit model must include the resistance and inductance of each winding. The circuit resistance is simply the sum of the forcing resistance (specified in the drive circuit design) and the winding resistance (specified by the manufacturer). In the hybrid stepping motor the inductance of the windings is almost independent of the rotor position and the average inductance specified by the manufacturer may be used. The phase circuit model shown in Fig.5.3 includes the total circuit resistance (R) and the average winding inductance (L).

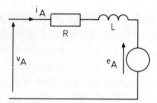

Fig. 5.3   *Circuit model for one phase of a hybrid motor*

The complete model must also take account of the voltages induced in the phase winding by rotor motion. These voltages occur because the permanent-magnet flux linking each winding varies sinusoidally with the position of the rotor. The flux linking phases A and B of a motor with p rotor teeth can be expressed as:

$$\psi_A = \psi_M \sin p\theta$$
$$\psi_B = \psi_M \sin(p\theta - \pi/2)$$
(5.1)

where ψ_M is the maximum flux linking each winding. When the rotor is moving at a speed $d\theta/dt$ the induced voltages in the phase windings are equal to the rate of change of flux linkages:

$$e_A = d\psi_A/dt = p\psi_M \cos(p\theta)\, d\theta/dt$$
$$e_B = d\psi_B/dt = p\psi_M \cos(p\theta - \pi/2)\, d\theta/dt$$
(5.2)

So the complete phase A circuit model includes the induced voltage e_A, which opposes the applied phase voltage v_A. Details of the induced voltage are not usually supplied by the stepping motor manufacturer, but fortunately the voltage can be measured experimentally. If the motor under test has its windings open-circuit and is driven by another motor (Fig.5.4) then the phase voltage is equal to the induced

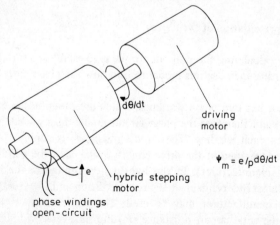

Fig. 5.4   *Determination of the magnet flux linked with the phase windings of a hybrid motor*

voltage, since the phase current is zero. Knowing the speed $(d\theta/dt)$, open-circuit voltage (e) and the number of rotor teeth (p), ψ_M can be determined from Eqn.(5.2). Alternatively ψ_M can be deduced from the pull-out torque of the motor at low speeds, as we shall see in the next section.

The hybrid stepping motor has two phases which are excited by positive or negative currents. A complete excitation cycle consists of four steps, corresponding to excitation of each phase by each current polarity. If the motor is operating at a stepping rate f then the excitation cycle repeats at a frequency $f/4$ and therefore the angular frequency of the excitation cycle is:

$$\omega = 2\pi \times (f/4) = \pi f/2 \tag{5.3}$$

During each excitation cycle the rotor moves one tooth pitch $(=2\pi/p)$ in a time $2\pi/\omega$, so the average rotor velocity is given by:

$$d\theta/dt = \text{distance/time}$$

$$= (2\pi/p)/(2\pi/\omega) = \omega/p \tag{5.4}$$

Integrating this equation with respect to time:

$$p\theta = \omega t - \delta \tag{5.5}$$

where δ is a constant of integration known as the 'load angle'. This angle accounts for the lag of the rotor behind the phase equilibrium position as the load on the motor increases.

The voltage applied to each phase circuit is a d.c. supply which can be switched on or off in the positive or negative sense. This switched supply introduces a non-linearity, which can be eliminated by considering only the fundamental components of voltage and current. In the next Section the torque produced by the motor is shown to depend on the interaction of the phase current and induced phase voltage. As the induced voltage is essentially sinusoidal only the sinusoidal component of phase current at the same frequency is required for the torque calculations. Substituting from Eqn. (5.5) into Eqn. (5.2) gives an expression for the variation of induced voltage:

$$e_A = \omega \psi_M \cos(\omega t - \delta) \tag{5.6}$$

The frequency of the induced voltage is equal to the frequency of the fundamental component of the supply voltage and so it is the fundamental component of phase current which is required for the torque calculations.

As the phase circuit model contains no other non-linearities, the fundamental component of phase current is produced entirely by the fundamental component of phase voltage. This component needs to be calculated according to the excitation scheme being used. In Fig. 5.5, for example, a two-phases-on scheme is shown and

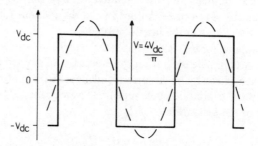

Fig. 5.5 *Phase voltage waveform for two-phases-on operation of a hybrid motor, showing the fundamental voltage component*

the fundamental voltage component has an amplitude of $V = 4V_{dc}/\pi$, where V_{dc} is the d.c. supply voltage. The fundamental components of phase voltage can be written as:

$$v_A = V \cos \omega t$$
$$v_B = V \cos(\omega t - \pi/2)$$

(5.7)

The instantaneous voltages and currents in phase A are related by the equation:

$$v_A = Ri_A + L(di_A/dt) + e_A$$

(5.8)

The fundamental current component in phase A can be expressed as

$$i_A = I \cos(\omega t - \delta - a)$$

where a is a phase angle. Substituting the fundamental voltage and current components into Eqn. (5.8) gives:

$$V \cos \omega t = RI \cos(\omega t - \delta - a) - \omega LI \sin(\omega t - \delta - a)$$
$$+ \omega \psi_M \cos(\omega t - \delta)$$

(5.9)

The phasor diagram corresponding to this equation is shown in Fig.5.6.

In the phasor diagram the applied phase voltage is equal to the vector sum of the induced voltage and the voltage drops across the resistance and inductance. The resistive voltage drop is in phase with the current, while the inductive voltage drop leads the current by $\pi/2$. The induced voltage leads the phase current by

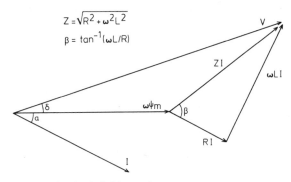

Fig. 5.6 *Phasor diagram for the hybrid stepping motor*

the angle a and lags the applied voltage by δ. It must be emphasised that this phasor diagram applies only to the fundamental current component and that the complete phase current waveform contains many other higher frequency components, which do not contribute to the torque produced by the motor.

5.2.2 Calculation of pull-out torque

Having established the phasor relationship between the fundamental components of phase voltage and current, the pull-out torque can be found directly. The mechanical output power per phase is simply the product of the phase current and the induced voltage (Fitzgerald and Kingsley, 1952):

Mechanical output power per phase

$$= e_A \, i_A$$

$$= \omega\psi_M \cos(\omega t - \delta) \times I \cos(\omega t - \delta - a)$$

$$= \frac{\omega\psi_M I \cos a}{2} + \frac{\omega\psi_M I \cos(2\omega t - 2\delta - a)}{2} \tag{5.10}$$

As the system is assumed to have an inertia large enough to prevent significant changes in velocity, the time-varying components of mechanical output power are of no interest and only the first term of Eqn. (5.10) need be evaluated. It is at this stage that the justification for neglecting harmonics of phase current becomes apparent. The expression $e_A \, i_A$ can only produce a constant term if both e_A and i_A have equal frequencies and, since e_A has only a fundamental component, it is the corresponding component of i_A which is required.

As the hybrid motor has two phases the total average mechanical output power is:

$$\text{Mechanical output power} = \omega\psi_M I \cos a \qquad (5.11)$$

The term $I \cos a$ is eliminated from this equation by constructing the line AA' in the phasor diagram (Fig. 5.7) at an angle a to the phasor ZI. Two expressions for

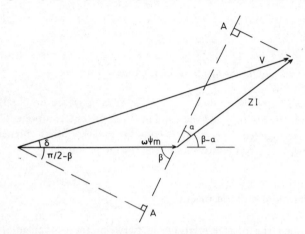

Fig. 5.7 *Phasor diagram construction to give an expression for I cos α*

the length of the line AA' are found by projecting the phasors $\omega\psi_M$, V and ZI onto the line:

$$AA' = V \sin(\pi/2 - \beta + \delta) = \omega\psi_M \cos \beta + ZI \cos \alpha$$

and so:

$$I \cos \alpha = [V \sin(\pi/2 - \beta + \delta) - \omega\psi_M \cos \beta]/Z$$

$$= [V \cos(\beta - \delta) - \omega\psi_M \cos \beta]/Z \qquad (5.12)$$

Substituting for $I \cos a$ from Eqn. (5.12) into Eqn. (5.11):

$$\text{Mechanical output power} = [\omega\psi_M V \cos(\beta - \delta) - \omega^2 \psi^2_M \cos \beta]/Z$$

but this mechanical output power is equal to the product of motor torque and speed. From Eqn. (5.4) the speed is ω/p and therefore:

$$\text{Torque} = \text{Power/Speed} = p[\psi_M V \cos(\beta - \delta) - \omega\psi^2_M \cos \beta]/Z \qquad (5.13)$$

An alternative approach (Hughes *et al.*, 1976) to this expression is to use the d-q axis transformation to solve the electrical equations of the system.

For a given set of motor and drive parameters at a fixed speed the only variable in Eqn. (5.13) is the load angle (δ), which varies according to the load so that the motor and load torques are equal. When the pull-out load is applied the load angle is such that the expression for torque is maximised. By inspection of Eqn. (5.13) the maximum torque occurs if $\beta = d$, so:

$$\text{Pull-out torque} = p[\psi_M V - \omega \psi_M^2 \cos \beta]/Z$$

$$= \frac{p\psi_M V}{\left(R^2 + \omega^2 L^2\right)^{1/2}} - \frac{p\omega\psi_M^2 R}{\left(R^2 + \omega^2 L^2\right)} \tag{5.14}$$

This apparently complicated expression for pull-out torque gives the surprisingly simple characteristic shown in Fig. 5.8.

At low speeds the pull-out torque is equal to $p\psi_M V/R$. Since this torque may be deduced from the static torque/rotor position characteristic, using the methods described in Chapter 4, the magnet flux (ψ_M) can be found in terms of the peak static torque.

As the stepping rate is increased the pull-out torque gradually reduces, until in some cases it reaches zero at a finite speed, the value ω_M shown in Fig. 5.8.

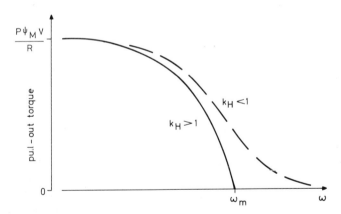

Fig. 5.8 *Alternative forms of a predicted pull-out torque/speed characteristic for the hybrid motor*

For other sets of motor/drive parameters the characteristic is asymptotic to the $T=0$ axis; the torque tends to zero at an infinite stepping rate. These two alternatives can be investigated quite simply by setting the pull-out torque expression (5.14) to zero and attempting to solve for the angular frequency, ω_m:

$$\frac{p\psi_M V}{\left(R^2 + \omega_m^2 L^2\right)^{1/2}} - \frac{p\omega_m \psi_M^2 R}{\left(R^2 + \omega_m^2 L^2\right)} = 0$$

and therefore:

$$\omega_m = R / L(k_H^2 - 1)^{1/2} \qquad\qquad [5.15(a)]$$

where $k_H = \psi_M R/VL$ is a constant for the motor. From Eqn.[5.15 (a)] we see that a real value for the maximum operating frequency can only be obtained if k_H is greater than unity, so that the square root in the denominator can be evaluated. The maximum stepping rate corresponding to Eqn. [5.15(b)] can be found from Eqn. (5.3):

$$f_m = 2\omega_m/\pi$$

The maximum operating frequency, given by Eqn. [5.15(a)] , is proportional to the total phase resistance (R), so if a motor is to operate at high speeds its drive circuit must include large forcing resistances. In choosing a motor for high speed operation a low value of k_H is desirable, as the denominator in the expression for maximum operating frequency is then minimised.

For values of k_H less than unity the pull-out torque/speed characteristic is asymptotic to the $T=0$ axis. The situation at high speeds can be investigated by letting ω tend towards infinity in Eqn. (5.14):

$$\text{Pull-out torque} = [p\psi_M V/ \omega L] - [p\psi_M^2 R / \omega L^2]$$

$$= (p\psi_M V/R) \times (R/\omega L) \times (1 - k_H) \qquad [5.15(b)]$$

The first factor is simply the pull-out torque of the motor at low speeds. The second factor shows that at the highest speeds the pull-out torque is inversely proportional to the supply frequency and that, as before, a large total phase resistance improves the high speed performance. Finally we see that the constant k_H is important; a motor with low k_H has more torque at high speeds.

It is interesting to note the physical significance of the parameter $k_H = \psi_M/(VL/R)$. The numerator of this expression, ψ_M, is the magnet flux linked with the phase winding. In the denominator the factor V/R is proportional to the winding current and therefore VL/R is the flux linked with the phase winding due to the current in the winding, so:

$$k_H = \frac{\text{Magnet flux linkages}}{\text{Winding self-flux linkages}}$$

From Eqn. [5.15(a)] and [5.15(b)] a low value of k_H improves the effective speed range of the motor, so if the motor is to operate at high speeds its permanent-magnet field must be weak compared to the field produced by the winding currents. If electromagnetic damping of the single-step response is required, however, the results of Section 4.3.3 indicate that a high value of k_H is needed; the choice of motor is then a compromise between the conflicting demands of damping and high-speed operation.

This analysis assumes that the voltage waveform applied to the phase circuit is independent of speed, but with a bridge drive circuit this may not be true. For two-phases-on excitation of the hybrid motor each phase is excited continuously by either positive or negative voltages. Fig. 5.9(a) shows typical current waveforms at

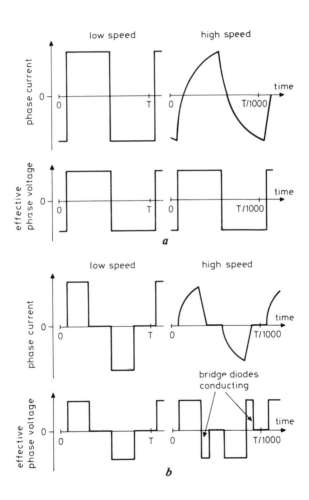

Fig. 5.9   *Phase current and voltage waveforms at low and high speed for a transistor bridge bipolar drive*
a   two-phases-on
b   one-phase-on

low and high speeds. When operating at low speed the freewheeling time of the phase currents is short compared to the total excitation time and for most of the cycle the phase current is carried by the switching transistors. At high speeds the freewheeling time is relatively long but the effective phase voltage is unchanged, even though the bridge diodes are conducting for a substantial part of the cycle.

With one-phase-on excitation, however, there are times in the excitation cycle when the phase voltage is zero. Fig. 5.9(b) shows that at low speeds the voltage is essentially as expected, but at high speeds the phase current is freewheeling through the bridge diodes against the supply voltage for a significant proportion of the cycle. During these freewheeling intervals the effective phase voltage is equal to the d.c. supply voltage. Therefore with one-phase-on operation of the hybrid motor the fundamental component of the phase voltage increases with speed and consequently the pull-out torque at high speeds is greater than that predicted by Eqn. (5.14) with V constant.

During deceleration the motor produces (negative) braking torque, which [from Eqn. (5.13) with $\beta - \delta = \pi$] has maximum value:

$$\text{Braking torque} = -\frac{p\psi_M V}{\left(R^2 + \omega^2 L^2\right)^{1/2}} - \frac{p\omega\psi_M^2 R}{\left(R^2 + \omega^2 L^2\right)} \tag{5.16}$$

Comparing Eqns. (5.14) and (5.16), we see that the two terms in Eqn. (5.14) are of opposite sign, whereas the terms in Eqn. (5.16) have the same sign, so the motor is able to produce more decelerating than accelerating torque at any stepping rate. This difference in torque capability also applies to variable-reluctance stepping motors and is very significant in determining optimum velocity profiles (Section 6.3).

Example

A 1·8 degree step-length hybrid motor has the following specification:

> *Rated phase current = 1·0 A*
> *Peak static torque = 0·8 Nm*
> *(one phase excited by rated current)*
> *Phase inductance = 10 mH*
> *Phase resistance = 2·0 ohms*

If the drive circuit incorporates a forcing resistance of 18·0 ohms and the motor is operated with two-phases-on excitation, find the pull-out torque at 500 steps per second and the maximum (no-load) stepping rate.

The solution to this problem is in two parts. Firstly we note that the magnet flux linkage with each winding is not quoted explicitly, so this parameter of the motor must be calculated from the information given. The pull-out torque and maximum stepping rate can then be found from Eqn. (5.14).

For a hybrid stepping motor with p rotor teeth the step length is given by Eqn. (1.3):

$$\text{Step length} = 90/p \text{ degrees}$$

so the motor with a 1·8 degree step length has 50 rotor teeth ($p=50$).

The pull-out torque at low speeds can be expressed in terms of the peak static torque using the methods described in Section 4.2. For two-phases-on excitation of the hybrid motor the peak static torque is 1·4 times the peak static torque with one-phase-on (Section 3.4.2), i.e. $T_{PK} = 1\cdot4 \times 0\cdot8$ Nm $= 1\cdot12$ Nm. The average torque produced by the motor over one step at maximum load is:

$$\text{Low-speed pull-out torque} = \frac{1}{\pi/2} \int_{\pi/4}^{3\pi/4} T_{PK} \sin (p\theta)d(p\theta)$$

$$= 2\sqrt{2}T_{PK}/\pi = 1\cdot02 \text{ Nm}$$

The total phase resistance, $R = 18\cdot0 + 2\cdot0$ ohms $= 20\cdot0$ ohms and for a rated phase current of 1·0 A the required d.c. supply voltage $V_{dc} = 20$ V. For two-phases-on excitation the fundamental component of supply voltage is $V = 4 \times V_{dc}/\pi$ (Fig. 5.5), i.e. $V = 25\cdot5$ V. The peak magnet flux linkages with each phase winding (ψ_M) can now be found by solving Eqn. (5.14) at low speeds ($\omega = 0$):

$$\text{Pull-out torque} = p\ \psi_M\ V\ /\ R$$

$$\therefore \psi_M = (\text{pull-out torque}) \times R\ /\ p \times V$$

$$= 1\cdot02 \times 20\cdot0/50 \times 25\cdot5 \text{ Wbt}$$

$$= 0\cdot016 \text{ Wbt}$$

At 500 steps per second the angular frequency of the supply [from Eqn. (5.3)] is:

$$\omega = n \times 500/2 \text{ rad s}^{-1}$$

$$= 785 \text{ rad s}^{-1}$$

The pull-out torque at 500 steps per second can be found by substituting into Eqn. (5.14):

$$V = 25 \cdot 5 \text{ V} \qquad R = 20 \text{ ohms}$$
$$L = 10 \text{ mH} \qquad \psi_M = 0 \cdot 016 \text{ Wbt}$$
$$p = 50 \qquad \omega = 785 \text{ rad s}^{-1}$$

$$\text{Pull-out torque} = 0 \cdot 53 \quad \text{Nm}$$

The maximum operating speed with zero load torque can be found by substituting into Eqn. (5.15):

$$k_H = \psi_M R / VL = 0 \cdot 016 \times 20/20 \times 0 \cdot 01 = 1 \cdot 6$$

$$\omega_m = R/L(k_H^2 - 1)^{1/2} = 20/0 \cdot 01 \times (1 \cdot 6^2 - 1)^{1/2} \text{ rad s}^{-1}$$

$$= 1600 \text{ rad s}^{-1}$$

This is the supply angular frequency at the maximum stepping rate, f_m, and from Eqn. (5.4):

$$\text{Maximum stepping rate}, f_m = 2 \times \omega_m/\pi = 1000 \text{ steps s}^{-1}$$

5.3   Pull-out torque/speed characteristics for the variable-reluctance motor

5.3.1   Circuit representation of the motor

In the variable-reluctance stepping motor the voltages induced by the rotor motion are due to the variation of the phase inductance with rotor position. If the stator and rotor teeth of one phase are fully aligned the flux path has a low reluctance; for a given phase current a large flux links the windings, so the phase inductance is at its maximum value. Conversely when the teeth are completely misaligned the flux path has a high reluctance and the phase inductance is a minimum. The variation of phase inductance is approximately sinusoidal with a wavelength equal to the rotor tooth pitch, as shown in Fig. 5.10. For a motor with p rotor teeth the phase inductance can be represented by:

$$L_A = L_0 + L_1 \sin p\theta \tag{5.17}$$

where L_0 is the average phase inductance and L_1 is the amplitude of the inductance variation with rotor position. These parameters may be specified by the manufacturer, but they can also be measured by the user if necessary (Kordik, 1975).

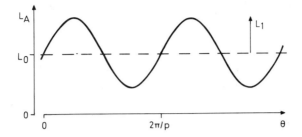

Fig. 5.10 *Variation of phase inductance with rotor position for a variable-reluctance stepping motor*

The complete excitation cycle for an *n*-phase motor consists of *n* steps, giving a total movement of one rotor tooth pitch $(2\pi/p)$. For a stepping rate *f*, the excitation frequency is f/n and therefore the angular frequency ω, of the supply to one phase is:

$$\omega = 2\pi f/n \tag{5.18}$$

The average rotor velocity over one supply cycle is:

$$d\theta/dt \ = \text{distance/time}$$

$$= (2\pi/p) \, / \, (2\pi/\omega) = \omega/p \tag{5.19}$$

Integrating Eqn. (5.19) with respect to time gives the variation of rotor position with time:

$$p\theta = \omega t - \delta \tag{5.20}$$

where δ is the load angle

For a phase current i_A, the flux linked with phase *A* is simply the product of current and inductance:

$$\psi_A = L_A i_A \tag{5.21}$$

and the rate of change of flux linkages with time is:

$$d\psi_A/dt \ = i_A \, dL_A/dt + L_A \, di_A/dt \tag{5.22}$$

$$= i_A \, (dL_A/d\theta) \times (d\theta/dt) + L_A \, di_A/dt$$

The first term in this expression is the voltage induced in the phase windings by the rotor motion and the second term arises from the changing current in the phase inductance. Substituting from Eqns. (5.17) and (5.20) gives the motional voltage in terms of the phase inductance:

$$e_A = i_A \ (dL_A/d\theta) \times (d\theta/dt)$$

$$= i_A \ (pL_1 \cos p\theta) \times (\omega/p)$$

$$= \omega L_1 i_A \cos (\omega t - \delta) \tag{5.23}$$

Since the phase current, i_A, is produced by a switched voltage supply it is non-sinusoidal and therefore, from Eqn. (5.23), the motional voltage is also non-sinusoidal. If the mechanical output power is to be found by evaluating the product of phase current and motional voltage, it appears that a large number of harmonics will need to be considered.

The analysis can be simplified by concentrating on the d.c. and fundamental components of voltage and current. By neglecting the harmonics of current an error is introduced into the analysis, but the next Section shows how this error can be corrected by evaluating the pull-out torque at low speeds. Using this approximation the phase voltage can be written:

$$v_A = V_0 + V_1 \cos \omega t \tag{5.24}$$

where V_0 is the d.c. component and V_1 is the fundamental component of the phase voltage. In Fig. 5.11, for example, the half-stepping excitation scheme is illustrated.

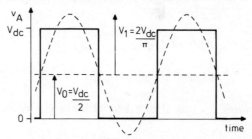

Fig. 5.11 *Fundamental and d.c. components of phase voltage for the half-stepping excitation scheme*

The phase is excited for half of the total cycle and therefore the d.c. component is half of the unipolar supply voltage ($V_0 = V_{dc}/2$) and the fundamental component is $V_1 = 2V_{dc}/\pi$.

Similarly the phase current can be approximated by its d.c. and fundamental components:

$$i_A = I_0 + I_1 \cos(\omega t - \delta - a) \tag{5.25}$$

where a is a phase angle. The relationship between the fundamental and d.c. components of current and voltage can be obtained from the phase voltage equation:

$$v_A = Ri_A + L_A(di_A/dt) + e_A$$

Substituting from Eqns. (5.17), (5.20), (5.23), (5.24), and (5.25) this voltage equation can be rewritten:

$$V_0 + V_1 \cos \omega t = RI_0 + RI_1 \cos(\omega t - \delta - a)$$
$$+ [L_0 + L_1 \sin(\omega t - \delta)] \times [-\omega I_1 \sin(\omega t - \delta - a)]$$
$$+ \omega L_1 \cos(\omega t - \delta) \times [I_0 + I_1 \cos(\omega t - \delta - a)]$$
$$= RI_0 + RI_1 \cos(\omega t - \delta - a) - \omega L_0 I_1 \sin(\omega t - \delta - a)$$
$$+ (\omega L_1 I_1/2) \times \cos(2\omega t - 2\delta - a) - (\omega L_1 I_1/2) \times \cos a$$
$$+ \omega L_1 I_0 \cos(\omega t - \delta) + (\omega L_1 I_1/2) \times \cos(2\omega t - 2\delta - a)$$
$$+ (\omega L_1 I_1/2) \times \cos a$$

Neglecting terms of greater than fundamental frequency and equating terms of d.c. and fundamental frequency:

$$V_0 = RI_0 \tag{5.26}$$

$$V_1 \cos \omega t = RI_1 \cos(\omega t - \delta - a) - \omega L_0 I_1 \sin(\omega t - \delta - a)$$
$$+ \omega L_1 I_0 \cos(\omega t - \delta) \tag{5.27}$$

Eqn. (5.26) contains no speed-dependent terms and therefore, as might be expected, the d.c. component of current is constant over the entire speed range. There is a broad similarity between the equation linking the fundamental voltage and current components in the variable-reluctance motor (5.27) and the phase voltage equation for the hybrid motor (5.9). In the variable-reluctance case the magnet flux (ψ_M) is replaced by the product ($L_1 I_0$) of the d.c. phase current and the variation of inductance with position. We may therefore proceed directly to the phasor diagram representation of Eqn. (5.27), as shown in Fig. 5.12. This phasor diagram is similar to Fig. 5.6 for the hybrid motor, except for minor changes of variable.

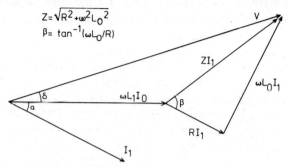

Fig. 5.12   *Phasor diagram for the variable-reluctance stepping motor*

5.3.2   Torque correction factor

Before proceeding to the discussion of pull-out torque/speed characteristics for the variable-reluctance motor, we must pause to consider in more detail the effects of neglecting the harmonics of phase current. In the previous Section it has been shown that these harmonics can contribute to the mechanical output power, because harmonics of current lead to harmonics of motional voltage. The approach adopted here is to introduce a 'torque correction factor' which allows for the harmonic contribution to the pull-out torque. It is assumed that the correction factor is independent of speed and therefore it can be evaluated quite simply for the rectangular current waveforms typical of low-speed operation.

The instantaneous torque produced by one phase of a variable-reluctance motor is, in the absence of magnetic saturation, proportional to the square of phase current and the rate of change of inductance with rotor position (Fitzgerald and Kingsley, 1952):

$$T_A = i_A^2 \, (dL_A/d\theta)/2 \tag{5.28}$$

If the phase current is approximated by its d.c. and fundamental components then substituting into Eqn. (5.28) from Eqns. (5.17), (5.20) and (5.25):

$$T_A = [I_0 + I_1 \cos(\omega t - \delta - a)]^2 \times [pL_1 \cos(\omega t - \delta)]/2$$

$$= [I_0^2 + 2I_0 I_1 \cos(\omega t - \delta - a) + I_1^2 \cos(\omega t - \delta - a)]$$

$$\times [pL_1 \cos(\omega t - \delta)]/2$$

The constant component of torque in this expression arises from the product of the ωt terms in the two factors:

$$T_{A\,const.} = pL_1 I_0 I_1 \cos(a)/2$$

and this has a maximum value when $a = 0$, giving a pull-out torque at low speeds:

$$T_{A\ pull\text{-}out} = pL_1I_0I_1/2$$

This is a contribution of the d.c. and fundamental current components to the pull-out torque. The magnitude of this contribution can be calculated for any excitation scheme by expressing the current components in terms of the rated winding current using a Fourier analysis of the waveform. This process is illustrated in Fig. 5.13 for the three-phase motor and the results are summarised in Table 5.1.

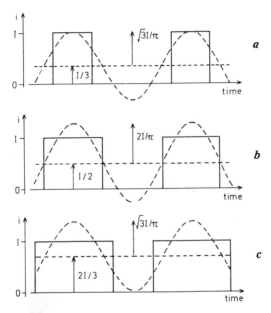

Fig. 5.13 *Fundamental and d.c. current components at low-speed for unipolar excitation of a three-phase motor*
a one-phase-on
b half-stepping
c two-phases-on

Table 5.1

Excitation scheme	I_0	I_1
One-phase-on	$I/3$	$\sqrt{3}I/\pi$
Half-stepping	$I/2$	$2I/\pi$
Two-phases-on	$2I/3$	$\sqrt{3}I/\pi$

I = Rated phase current = V_{dc}/R

The total torque, when all current components are considered, can be found by evaluating Eqn. (5.28) for the rectangular current waveform using a graphical method. For maximum (pull-out) torque the i_A^2 and $dL_A/d\theta$ waveforms must have the correct phase relationship, as shown in Fig. 5.14. In this example the half-

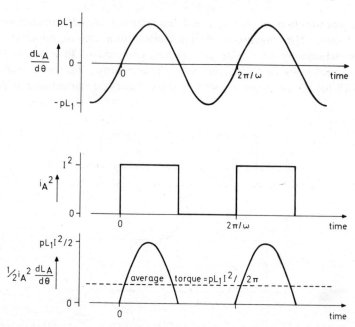

Fig. 5.14   *Graphical method of determining pull-out torque at low speeds for half-stepping unipolar excitation*

stepping excitation scheme is being used, so the phase is excited for half of the total cycle and the torque is maximised if the phase current is turned on when $dL_A/d\theta$ is positive. The average torque per phase can then be found by averaging the instantaneous torque over one cycle. This method can be repeated for each excitation scheme to give an exact expression for the pull-out torque in terms of phase current and inductance. In Table 5.2 the exact graphical results are compared with those obtained by evaluating Eqn. (5.29) with the current components given in Table 5.1.

Table 5.2

Excitation scheme	Pull-out torque		Correction factor
	Approximate	Exact	
One-phase-on	$pL_1I^2/\pi2\sqrt{3}$	$\sqrt{3}pL_1I^2/4\pi$	1·50
Half-stepping	$pL_1I^2/2\pi$	$pL_1I^2/2\pi$	1·00
Two-phases-on	$pL_1I^2/\pi\sqrt{3}$	$\sqrt{3}pL_1I_2/4\pi$	0·75

The torque correction factor shown in the final column of Table 5.2 is the ratio of the exact to approximate results for the pull-out torque. With the half-stepping excitation scheme, for example, the pull-out torque is predicted precisely from the d.c. and fundamental current components. If one-phase-on excitation is used, however, the pull-out torque obtained from the d.c. and fundamental components must be multiplied by 1·5 to give the exact pull-out torque. Comparing one- and two-phase-on excitation schemes it is perhaps surprising to find that the exact analysis shows that the torque produced is the same, whereas the approximate result indicates that the two-phases-on scheme should produce twice as much torque as one-phase-on excitation. However in Chapter 3 we have already seen that the peak static torque of a three-phase motor is the same for one- and two-phases-on and this result therefore extends to the pull-out torque at low speeds.

Having established the torque correction factors for the common excitation schemes of a three-phase motor and illustrated the method by which the factor may be found for motors with larger numbers of phases, we can now consider how the pull-out torque produced by the d.c. and fundamental current components varies with stepping rate.

5.3.3   Calculation of pull-out torque

The similarity between the phasor diagrams for the fundamental current components in the hybrid (Fig. 5.6) and variable-reluctance (Fig. 5.12) stepping motors has already been noted. By analogy to the hybrid motor the expression for pull-out torque in a variable-reluctance motor can be found directly. Writing Eqn. (5.14) with $\psi_M = L_1 I_0$, $V = V_1$, $L = L_0$:

$$\text{Pull-out torque} = \frac{n}{2}\left[\frac{pL_1 I_0 V_1}{(R^2 + \omega^2 L_0^2)^{1/2}} - \frac{p\omega L_1^2 I_0^2 R}{R^2 + \omega^2 L_0^2}\right] \tag{5.30}$$

The multiplier $n/2$ takes account of the difference in the number of phases between the two types. The hybrid motor has two phases, so the pull-out torque per phase is $1/2$ above the value indicated in Eqn. (5.14). The variable-reluctance motor has n phases, so the pull-out torque per phase must be multiplied by n, with the overall result shown in Eqn. (5.30).

Typical pull-out torque/speed characteristics for the variable-reluctance stepping motor are shown in Fig. 5.15. An expression for the maximum operating speed of the variable-reluctance motor follows from Eqn. [5.15(a)], using the same substitutions as above:

$$\omega_M = R/L_0(k_v^2 - 1)^{1/2}$$

if $k_v > 1$. The corresponding maximum stepping rate can be found from Eqn.

(5.18):

$$f_m = \omega_m n/2\pi = nR/2\pi L_0 (k_v^2 - 1)^{1/2} \qquad [5.31(a)]$$

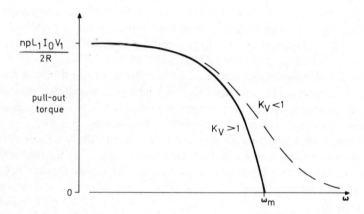

Fig. 5.15   *Alternative forms of the predicted pull-out torque/speed characteristic for the variable-reluctance motor*

For values of $k_v < 1$ the pull-out torque/speed characteristic is asymptotic to the $T=0$ axis and the parallel expression to Eqn. [5.15(b)] gives the pull-out torque of the variable-reluctance motor at high speeds:

$$\text{Pull-out torque} = (npL_1 I_0 V_1/2R) \times (R/\omega L_0) \times (1 - k_v) \qquad [5.31(b)]$$

The motor parameter k_v is given by:

$$k_v = L_1 I_0/(V_1 L_0/R) = L_1 V_0/L_0 V_1 \qquad (5.32)$$

The parameter k_v depends on the ratio of the motor inductances (L_1/L_0) and on the ratio of the d.c. to fundamental voltage components (V_0/V_1), which is a function of the excitation scheme. Eqns. (5.31) indicate that a low value of k_v is required if the motor is to operate over a wide speed range. As far as the choice of motor is concerned Eqn. (5.32) indicates that the variation of inductance with rotor position (L_1) should be small compared to the average phase inductance (L_0). The speed range can also be improved by careful selection of the excitation scheme, so as to minimise the ratio V_0/V_1. For a three-phase motor the voltage component ratio for each excitation scheme is shown in Table 5.3:

Table 5.3

Excitation scheme	V_0/V_1
One-phase-on	0·60
Half-stepping	0·79
Two-phases-on	1·20

By operating the motor with one-phase-on excitation the speed range is considerably improved compared to two-phases-on operation, for which the value of k_v is doubled. As with the hybrid motor, there is a conflict between the design of a system for high-speed operation and for a well-damped response, because if one-phase-on excitation is used the single-step response is poorly damped. The half-stepping scheme is therefore a useful compromise, since the V_0/V_1 ratio is not too high and reasonable damping can be achieved by arranging for the motor to come to rest at positions where two phases are excited.

Eqns. (5.31) and (5.15) show that the performance of both variable-reluctance and hybrid stepping motors at high speeds is proportional to the total phase resistance (R), which can be adjusted by changing the series forcing resistance in the drive circuit. Obviously the forcing resistance is one of the fundamental factors influencing high-speed performance so this relationship is given further consideration in the next section.

Example

A three-phase variable-reluctance stepping motor has L_0 = 40 mH, L_1 = 20 mH, winding resistance 1 ohm and eight rotor teeth. Find the pull-out torque at 500 steps s^{-1} if the motor is operated with the one-phase-on excitation scheme and the drive circuit has supply voltage 30 V and forcing resistance 9 ohms.

From Eqn. (5.18):

$$\omega = 2\pi f/n = 2\pi \times 500/3 \text{ radians s}^{-1} = 1046 \text{ radians s}^{-1}$$

The components of the phase voltage (V_0 and V_1) have to be calculated for the one-phase-on excitation scheme. For 1/3 of the cycle the phase is excited by the supply voltage V_{dc}:

$$v = \begin{cases} 0 & -\pi < \omega t < -\pi/3 \\ V_{dc} & -\pi/3 < \omega t < \pi/3 \\ 0 & \pi/3 < \omega t < \pi \end{cases}$$

The d.c. component (V_0) is simply the average of the phase voltage:

$$V_0 = \frac{1}{2\pi} \int_{-\pi}^{\pi} v \, d(\omega t) = \frac{1}{2\pi} \int_{-\pi/3}^{\pi/3} V_{dc} \, d(\omega t) = V_{dc}/3$$

In this example $V_{dc} = 30$ V and therefore $V_0 = 10$ V.

The fundamental component is given by:

$$V_1 = \frac{1}{\pi} \int_{-\pi}^{\pi} v \cos(\omega t) \, d(\omega t) = \frac{1}{\pi} \int_{-\pi/3}^{\pi/3} V_{dc} \cos(\omega t) \, d(\omega t)$$

$$= \sqrt{3} V_{dc}/\pi$$

and so with $V_{dc} = 30$ V, $V_1 = 16 \cdot 5$ V.

In Eqn. (5.30) the motor/drive parameters are:

$$n = 3 \quad p = 8 \quad R = 9 + 1 \text{ ohms} = 10 \text{ ohms} \quad I_0 = V_0/R = 10/10 \text{ A} = 1 \cdot 0 \text{ A}$$

$$V_1 = 16 \cdot 5 \text{ V} \quad L_0 = 40 \text{ mH} \quad L_1 = 20 \text{ mH} \quad \omega = 1046 \text{ rads s}^{-1}$$

and the pull-out torque from the d.c. and fundamental current components is therefore 0·065 Nm

Finally the relevant torque correction factor is found from Table 5.2. For the one-phase-on excitation scheme the correction factor is 1·5, which is used to multiply the approximate result, so:

Pull-out torque at 500 steps s^{-1} = $1 \cdot 5 \times 0 \cdot 065$ Nm = $0 \cdot 098$ Nm

5.4   Drive circuit design

5.4.1   Drive requirements

The operating speed range of both hybrid and variable-reluctance stepping motors is proportional to the phase resistance [Eqns. (5.15) and (5.31)]. As the phase resistance can be controlled by changing the forcing resistance, it is possible to operate stepping motors at very high speeds using the simple drive circuits (described in Chapter 2) with a large forcing resistance. However the supply voltage must also be increased to maintain the phase current at its rated value when the motor is stationary and consequently a large d.c. power supply is needed. For small motors this may be a perfectly satisfactory method of obtaining a wide speed range, because the size of power supply is unimportant. With larger motors, however, the power supply may have to have a capacity of several kilowatts if the system is to operate over a satisfactory speed range. In these circumstances it is worth reconsidering the design of the drive circuit.

A large supply voltage and phase resistance are only required when the motor is operating at high speeds. If the motor is stationary the phase currents dissipate a substantial part of the supply output in the series forcing resistance and the heat produced can cause problems if the forcing resistances cannot be cooled. The simple series forcing resistance is therefore an inefficient method of improving the speed range; power is wasted in the resistances at low speeds so that the mechanical output power (torque x speed) can be improved at higher speeds.

An alternative viewpoint is obtained from the circuit model for the phase windings. For both types of motor the circuit model (Fig. 5.16) includes an

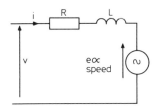

Fig. 5.16   *Circuit model for one phase*

induced voltage which is proportional to speed and the phase mechanical output power is the product of this voltage and the current flowing against it. Current flows into the motor provided the voltage applied to the phase can overcome the induced voltage. If the motor is to operate at a higher speed then the induced voltage is larger and the applied voltage must also increase, so that current flows into the winding over the extended speed range. Increases in applied voltage must be accompanied by proportional increases in phase resistance if the winding current is to be limited to its rated value when the motor is stationary. This argument reveals that it is the phase voltage which is the fundamental factor in determining the speed range of the motor and the function of the series resistance can then be regarded as 'current-limiting', rather than 'forcing'. At the highest speeds the phase current is low, so the voltage drop across the series resistance is small and the applied voltage balances the induced voltage.

The drive circuit requirements are now clarified: a large supply voltage is needed at high speeds, but the phase current at low speeds must be limited without the power wastage associated with the simple series resistance method of current-limiting. A considerable number of drive circuit configurations are available, but in the following Sections the discussion is limited to two of the best-known types.

5.4.2   Bilevel drive

In the bilevel drive there are two supply voltages. A high voltage is used when the phase current is to be turned on or off, while a lower voltage maintains the current at its rated value during continuous excitation..

The circuit diagram for one phase of a unipolar bilevel drive is shown in Fig. 5.17. When the winding is to be excited both transistors (T1 and T2) are

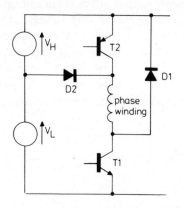

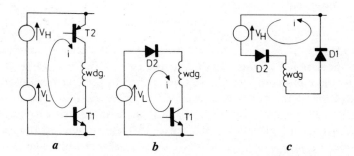

a	*b*	*c*

Fig. 5.17   *The bilevel drive and the effective circuits during the excitation interval*
    a   at turn-on
    b   continuous excitation
    c   at turn-off

switched on, so the voltage applied to the phase winding is equal to the sum of the two supply voltages $(V_L + V_H)$, the diode D2 being reverse-biased by V_H. There is no series resistance to limit the curent, which therefore starts to rise towards a value which is many times the rated winding current. After a short time, however, transistor T2 is switched off and the winding current flows from the supply voltage V_L via diode D2 and transistor T1. The rated winding current is maintained by the voltage V_L, which is chosen so that $V_L/R=$ rated current. At the end of the phase excitation interval transistor T1 is also switched off and the winding current is left to flow around the path through diodes D1 and D2. Rapid decay of the current is assured, because the high supply voltage V_H is included in this freewheeling path.

A typical current waveform for one excitation interval is illustrated in Fig. 5.18.

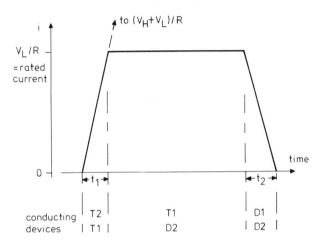

Fig. 5.18 *Phase current waveform for a bilevel drive*

Using a simple 'static inductance' model of the phase winding (i.e. neglecting the voltage induced by rotor motion), the times for current rise (t_1) and decay (t_2) can be calculated. If the winding has an effective inductance L and resistance R, then at turn-on the phase current is:

$$i = (V_H + V_L) \times [1 - \exp(-tR/L)]/R$$

and if $i \ll (V_H + V_L)/R$:

$$i = (V_H + V_L) \times t/L \tag{5.33}$$

Transistor T2 must remain switched on until the phase current reaches its rated value (V_L/R) at time t_1:

$$V_L/R = (V_H + V_L) \times t_1/L$$

$$t_1 = [V_L/(V_H + V_L)] \times (L/R) \tag{5.34}$$

The ratio $D = V_H/V_L$ is known as the 'overdrive' available in the circuit and Eqn.

(5.34) may be expressed in terms of D:

$$t_1 = (L/R)/(D+1) \tag{5.35}$$

If the overdrive is large the phase current is established more quickly and higher operating speeds are possible.

At a time t after T1 is turned off the phase current is:

$$i = -(V_H/R) + [(V_L + V_H)/R] \times \exp(-tR/L)$$

and if $i \ll V_H/R$:

$$i = V_L/R - (V_L + V_H)t/L \tag{5.36}$$

The time (t_2) taken for the current to decay to zero is given by:

$$0 = -(V_H + V_L)t_2/L$$

$$t_2 = (L/R)/(D+1) \tag{5.37}$$

Comparing Eqns. (5.35) and (5.37), the current decay time is equal to the current rise time.

Example

A bifilar-wound hybrid stepping motor has a winding inductance of 10 mH and a resistance of 2·0 ohms. If the rated winding current is 3 A, find the voltages required in a bilevel drive circuit to give a current rise time of 1 ms.

The low supply voltage must maintain the rated winding current when the motor is stationary:

$$\underline{V_L = R{\times}I = 2{\cdot}0{\times}3{\cdot}0 \text{ V} = 6{\cdot}0 \text{ V}}$$

The winding time constant $L/R = 10/2{\cdot}0$ ms $= 5{\cdot}0$ ms and the current rise time required is 1 ms. Substituting in Eqn. (5.35):

$$1 \text{ ms} = 5{\cdot}0/(D+1) \text{ ms}$$

and therefore $D = 4 = V_H/V_L$.

$$\underline{\text{High-voltage supply, } V_H = 4{\times}V_L = 24{\cdot}0 \text{ V}}$$

The bilevel drive has the merit of simplicity, because the only control circuitry required is concerned with the switching time of transistor T2 at the beginning of each excitation interval. As this transistor conducts for a fixed time, dictated by the winding time constant, it may be switched from a fixed-period monostable circuit triggered by the phase excitation signal.

One disadvantage of the bilevel drive in this form is that it is unable to counteract the motional voltage and this voltage has been neglected in the analysis of drive circuit performance. Once the winding current is established only the low-voltage supply is effective and this may be insufficient to overcome the motional voltage during the remainder of the excitation interval.

5.4.3   Chopper drive

This drive circuit – illustrated in its unipolar form in Fig. 5.19 – has a high supply

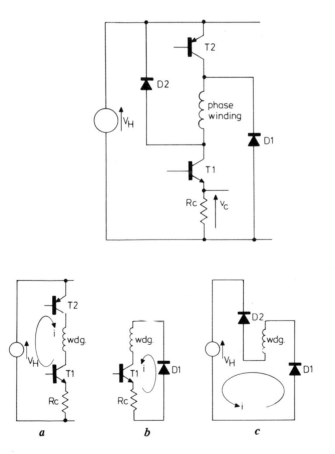

Fig. 5.19   *The chopper drive and the effective circuits during the excitation interval*
a   current less than rated
b   current greater than rated
c   at turn-off

voltage which is applied to the phase winding whenever the current falls below its rated value. If the phase excitation signal is present, the base drive for transistor T2 is controlled by the voltage v_c dropped across the small resistance R_c by the winding curent. At the beginning of the excitation interval the transistor T1 is switched on and the base drive to T2 is enabled. As the phase current is initially zero there is no voltage across v_c and the transistor T2 is switched on. The full supply voltage is therefore applied to the phase winding, as shown in the timing diagram, Fig. 5.20.

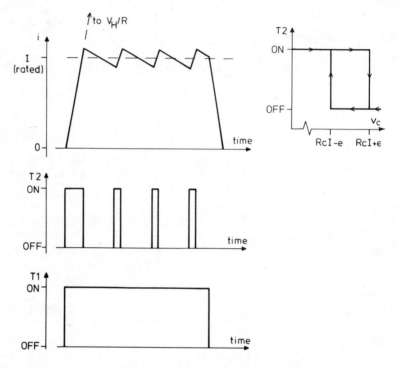

Fig. 5.20   *Chopper drive current waveform and transistor switching times*

The phase current rises rapidly until it slightly exceeds its rated value (I). Consequently the control voltage is $R_c I + e$ and this is sufficient to switch off transistor T2. There is now no voltage applied to the phase winding and the current decays around a path which includes T1, R_c and diode D1. This current path has a small resistance and no opposing voltage, so the decay of current is relatively slow. As the resistance R_c is still included in the circuit the winding current can be monitored and when the control voltage has fallen to $R_c I - e$ the transistor T2 is switched on again. The full supply voltage is applied to the winding and the current is rapidly boosted to slightly above rated. This cycle is repeated throughout the excitation time, with the winding current maintained near its rated value by an 'on-off' closed-loop control.

At the end of the excitation interval both transistors are switched off and the winding current freewheels via diodes D1 and D2. The current is now opposed by the supply voltage and is rapidly forced to zero. A high proportion of the energy stored in the winding inductance at turn-off is returned to the supply and therefore the system has a high efficiency.

The chopper drive incorporates more sophisticated control circuitry, e.g. the T2 base drive requires a Schmitt triggering of the control voltage v_c to produce the transition levels. If these levels are not well-separated the transistor T2 switches on and off at a very high frequency, causing interference with adjacent equipment and additional iron losses in the motor. However the chopper drive does have the advantage that the available supply voltage is fully utilised, enabling operation over the widest possible speed range, and the power losses in forcing resistors are eliminated, giving a good system efficiency.

Chapter 6

Open-loop control

6.1 Introduction

The initial stages of system design are concerned with steady-state performance; the choice of stepping motor and drive circuit is mainly dictated by the maximum tolerable position error and the maximum required stepping rate. When this task of selection is completed the designer must then consider how the motor and drive are to be controlled and interfaced to the remainder of the system. The aim of the remaining Chapters, therefore, is to show that system performance can be maximised and costs minimised by correct choice of control scheme and interfacing technique.

The open-loop control schemes discussed in this Chapter have the merits of simplicity and consequent low cost. A block diagram for a typical open-loop control system is shown in Fig. 6.1. Digital phase control signals are generated by

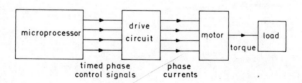

Fig. 6.1 *A microprocessor-based open-loop control*

the microprocessor and amplified by the drive circuit before being applied to the motor. Although the system illustrated receives its phase control signals from a microprocessor a number of less sophisticated alternatives are presented later in this Chapter.

Whatever the signal source, the designer needs to know what restrictions are imposed on the timing of the control signals by the parameters of the drive, motor and load. Some of these restrictions stem from the steady-state performance (e.g. the maximum stepping rate with a given load torque can be deduced from the pull-out torque/speed characteristics), but still more restrictions arise when transient performance is considered. If the system has a high inertia, for example, the maximum stepping rate cannot be attained instantaneously; the stepping rate must be gradually increased towards the maximum value so that the motor has sufficient time to accelerate the load inertia.

In an open-loop control scheme there is no feedback of load position to the controller and therefore it is imperative that the motor responds correctly to each excitation change. If the excitation changes are made too quickly the motor is unable to move the motor to the new demanded position and consequently there is a permanent error in the actual load position compared to the position expected by the controller. The timing of phase control signals for optimum open-loop performance is reasonably straightforward if the load parameters are substantially constant with time. However in applications where the load is likely to fluctuate the timings must be set for the worst conditions (i.e. largest load) and the control scheme is then non-optimal for all other loads.

6.2   Starting/stopping rate

The simplest form of open-loop control (Fig. 6.2) is a constant stepping rate which

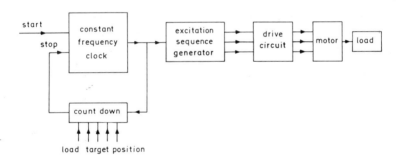

Fig. 6.2   *Constant stepping rate open-loop control*

is applied to the motor until the load reaches the target position. The sequence generator produces the phase control signals and is triggered by step command pulses from a constant frequency clock. This clock can be turned on by the START signal, causing the motor to run at a stepping rate equal to the clock frequency, and turned off by the STOP signal, in which case the motor is halted. Initially the target direction is sent to the phase sequence generator, which then produces phase control signals to turn the motor in the appropriate direction. The target position is loaded into a downcounter, which keeps a tally of the steps executed. Clock pulses are fed to both the phase sequence generator and the downcounter. Changes in phase excitation are therefore made at the constant clock frequency and the instantaneous position of the motor relative to the target is recorded in the downcounter. When the load reaches the target the downcounter contents are zero and this zero count is used to generate the clock STOP signal.

If the constant clock frequency is set too high the motor is unable to accelerate the load inertia to the corresponding stepping rate and the system either fails to operate at all or loses steps at the start of its travel. The maximum demanded stepping rate to which the motor can respond without loss of steps when initially at rest is known as the 'starting rate' or the 'pull-in rate'. Similarly the 'stopping rate' is the maximum stepping rate which can be suddenly switched off without the motor overshooting the target position. For any motor/load combination there is very little difference between the starting and stopping rates; viscous friction tends to reduce acceleration and starting rate, but aids retardation and therefore improves the stopping rate. In a simple constant frequency system, however, the clock must be set at the lower of the two rates to ensure reliable starting and stopping.

A rigorous analysis of acceleration from rest is possible (Pickup and Tipping 1976), but in practice a simplified approach (Gupta and Mathur, 1976); Lawrenson *et al.*, 1977) yields approximate results, which can be confirmed experimentally on a prototype system. Until a prototype is produced it is, in any case, not usually possible to define the load parameters to sufficient accuracy to justify a more detailed study.

When the motor is accelerating from rest the stepping rates are low; the period of each phase excitation interval is much longer than the electrical time constant of the phase circuit. In this situation the performance of the system can be predicted in terms of the motor's static torque/rotor position characteristic. As an example consider a four-phase motor with two-phases-on excitation, giving the approximately sinusoidal torque/position characteristics illustrated in Fig. 6.3. The load

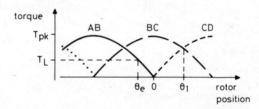

Fig. 6.3   *Static torque/rotor position characteristics for a four-phase motor*

torque is T_L and the peak static torque is T_{PK}, so if the motor is initially at rest with phases A and B excited the static position error (θ_e) is given by Eqn. (3.2):

$$\theta_e = \frac{\sin^{-1}(-T_L/T_{PK})}{p}$$

(6.1)

The first step command changes the excitation to phases B and C and the static torque at the position θ_e then exceeds the load torque, so the motor accelerates in the positive direction. During the first excitation interval (i.e. the time between first and second step commands) the motor must move far enough and attain sufficient velocity to ensure that synchronism with the step commands is maintained when the excitation changes from *BC* to *CD*. However there is no need for the motor to travel a complete step in this first interval, because a small lag behind the phase equilibrium position can be recovered over subsequent steps. A suitable (although rather conservative) estimate is that the motor should at least move to the 'cross-over' position of the *BC* and *CD* torque characteristics (θ_1), so that when phases *CD* are excited there is a high torque available to continue acceleration. In these circumstances the average torque produced by the motor while it is moving from position θ_e to θ_1 with phases *BC* excited is:

$$T_M = \frac{1}{\theta_1 - \theta_e} \int_{\theta_e}^{\theta_1} -T_{PK} \sin(p\theta - \pi/2)\, d\theta$$

$$= \frac{-T_{PK}}{p(\theta_1 - \theta_e)} [\cos(p\theta_1 - \pi/2) - \cos(p\theta_e - \pi/2)]$$

$$= \frac{T_{PK}}{p(\theta_1 - \theta_e)} [\sin(p\theta_1) - \sin(p\theta_e)] \tag{6.2}$$

The problem can now be simplified by assuming that the torque over the excitation interval is effectively constant and equal to this average value. Fig. 6.3 confirms that this assumption is unlikely to introduce any major errors; the instantaneous torque varies from T_{PK} (at $\theta=0$) to $0.7 T_{PK}$ (at $\theta=\theta_1$), so the maximum error is approximately 15%. The equation of motion for the system inertia (J) is:

$$T_M - T_L = J\, d^2\theta/dt^2 \tag{6.3}$$

Integrating Eqn. (6.3) twice with respect to time and using the initial conditions that at $t = 0$, $\theta = \theta_e$ and $d\theta/dt = 0$:

$$\theta = (T_M - T_L)t^2/J + \theta_e \tag{6.4}$$

After one period of excitation, t_p, the rotor is at the position θ_1, therefore:

$$\theta_1 = (T_M - T_L)t_p^2/J + \theta_e$$

$$t_p = [J(\theta_1 - \theta_e)/(T_M - T_L)]^{1/2}$$

so the starting rate for the four-phase motor is approximately:

$$\text{Starting rate} = 1/t_p = [(T_M - T_L)/J(\theta_1 - \theta_e)]^{1/2} \tag{6.5}$$

As might be expected, the starting rate is improved if the motor has a high torque (T_M) or the load torque (T_L) is low. Any reduction in system inertia (motor inertia + load inertia) also improves the starting rate.

If there is any doubt about the chosen criterion for distance moved during the first step then the second step can also be analysed to ensure that the motor maintains synchronism during that excitation interval. The average torque during the interval can be found from the static torque/rotor position characteristic. Using this value of torque, the equation of motion can be solved with the appropriate initial conditions: at $t = t_p$, $\theta = \theta_1$ and $d\theta/dt = (T_M - T_L)t_p/J$.

This simplified approach to starting rate can be applied to motors with any step length, excitation scheme and load. In cases where the torque/position is notably non-sinusoidal the average torque can be calculated using graphical methods, as illustrated in the following example.

Example

A three-phase variable-reluctance motor has a step length of 15 degrees and has one phase excited by the rated current. The corresponding torque/position character-istic is shown in Fig. 6.4. Estimate the starting rate for a load inertia of 2×10^{-4} kg m^2 and a load torque of 0·1 Nm.

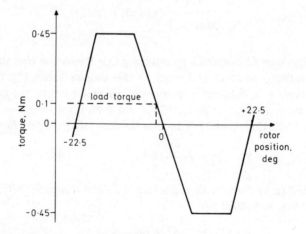

Fig. 6.4 *Static torque/rotor position characteristic*

The first stage is to draw the static torque/rotor position characteristics for each of the three phases. These characteristics are displaced from each other by the step length, as shown in Fig. 6.5.

The static position error can be deduced from the torque/position characteristic by drawing the line corresponding to a torque of 0·1 Nm. This line intersects the characteristic about 2 degrees from the equilibrium position, so $\theta_e = -2$ degrees $= -0.035$ radians. The 'crossover' point of the characteristics for phases B and C is at the position $\theta_1 = 10.5$ degrees $= 0.117$ radians.

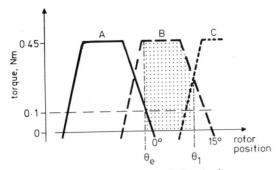

Fig. 6.5 *Acceleration from rest: average torque during the first step*

The average motor torque as the rotor moves between θ_e and θ_1 is the shaded area in Fig. 6.5 divided by the distance moved. As the boundaries of the relevant area are straight lines it can be evaluated quite simply:

Area $= (0{\cdot}45 \times 9{\cdot}5) + [(0{\cdot}45 + 0{\cdot}27) \times 3/2]$ Nm degrees

$= 53{\cdot}5$ Nm degrees

and so the average torque can be found from:

T_M = Area/distance = $53{\cdot}5/(10{\cdot}5 + 2{\cdot}0)$ Nm = $0{\cdot}428$ Nm

Substituting into Eqn. 6.5, with θ_1 and θ_e in radians:

Starting rate = $[(0{\cdot}428 - 0{\cdot}10)/2 \times (0{\cdot}117 + 0{\cdot}035)]^{1/2} \times 10^2$ steps s^{-1}

$= 104$ steps s^{-1}

To allow for slight changes in the load torque, due to wear of the components during the system's working life, the constant frequency clock would be set between 90 and 100 steps s^{-1}. The pull-out torque/speed characteristic would also be consulted to ensure that the system is not susceptible to mechanical resonances at the working speed. If the calculated starting rate does coincide with a resonant rate the designer can either elect to use a lower frequency clock or try to reduce the resonance with additional damping.

6.3 Acceleration/deceleration capability

In general the starting rate of a stepping motor system is much lower than its pull-out rate, so positioning times can be reduced substantially by continuing to accelerate the motor over several steps until the pull-out rate is attained. As the target position is approached the stepping rate is gradually reduced to the starting/stopping rate, so that the motor can be halted when the final position is reached. A graph of the stepping rate against time as the motor moves between the initial and target positions is commonly referred to as the 'velocity profile'; a typical example is shown in Fig. 6.6(*a*). Note that the rate of deceleration can be significantly faster

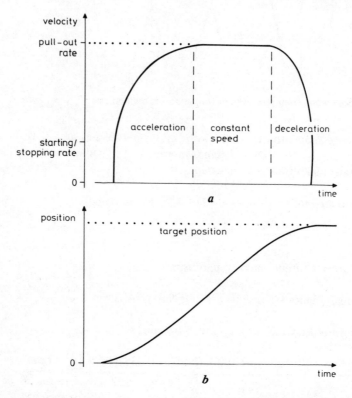

Fig. 6.6 *Acceleration to the pull-out rate and deceleration to the target position*
a velocity profile
b corresponding position/time response

than the acceleration, because the load torque tends to retard the system and the motor is able to develop more decelerating than accelerating torque (Section 5.2.2).

Various methods of generating the velocity profile, or approximations to it, are considered later in this Chapter, but firstly the relationship between the system parameters and acceleration/deceleration capability must be established. More sophisticated methods of open-loop control enable the system to approach its pull-out rate and therefore the dependence of motor torque on stepping rate must be taken into account. This is accomplished quite easily by using the pull-out torque/speed characteristic of the motor/drive as the basis for the analysis. Variations of load and friction torques with speed can also be taken into account.

Typical pull-out torque and load torque characteristics are shown in Fig. 6.7(*a*).

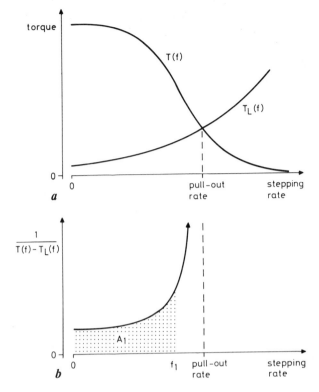

Fig. 6.7   *Derivation of the velocity profile*
a   pull-out torque $T(f)$ and load torque $T_L(f)$ characteristics
b $1/[T(f) - T_L(f)]$

In this case a linear speed scale must be used and therefore the shape of the characteristic is rather different from those illustrated in Chapters 4 and 5, where a logarithmic speed scale has been used. At a stepping rate f the pull-out torque is denoted by $T(f)$ and the load torque by $T_L(f)$. If the motor is to accelerate as quickly as possible the maximum (pull-out) torque must be developed at all speeds. This torque overcomes the load torque and accelerates the system inertia or, expressed algebraically:

$$T(f) = T_L(f) + J(d^2\theta/dt^2) \tag{6.6}$$

For a motor with n phases and p rotor teeth the step length is $2\pi/np$ and so the stepping rate is related to the rotor velocity by:

$$d\theta/dt = 2\pi f/np \tag{6.7}$$

Substituting Eqn. (6.7) into Eqn. (6.6):

$$T(f) = T_L(f) + (2\pi J/np) \times (df/dt)$$

$$df/dt = [T(f) - T_L(f)] \times np/(2\pi J)$$

This equation can be integrated to give the time, t, taken to reach the stepping rate, f, as the motor accelerates from rest:

$$\frac{np}{(2\pi J)} \int_0^t dt = \frac{npt}{2\pi J} = \int_0^f \frac{df}{T(f) - T_L(f)} \qquad (6.8)$$

In general this integral must be performed graphically, as both $T(f)$ and $T_L(f)$ are non-analytic functions. Fig. 6.7(b) shows the function $1/[T(f) - T_L(f)]$ and the shaded area, A_1, corresponds to the integral of this function with respect to stepping rate for rates between 0 and f_1. The time, t_1, taken to reach stepping rate f_1 can then be found from Eqn. (6.8):

$$t_1 = 2\pi J A_1/np \qquad (6.9)$$

A complete velocity profile for the acceleration can be built-up by repeating this process for a range of stepping rates up to the pull-out rate f_m. The procedure can be simplified if $T(f)$ and $T_L(f)$ can be approximated by analytic functions, as in the following example.

Example

A stepping motor has a step length of 15 degrees and the pull-out torque/speed characteristic shown in Fig. 6.8. It is used to drive a load with an inertia of 2×10^{-4}

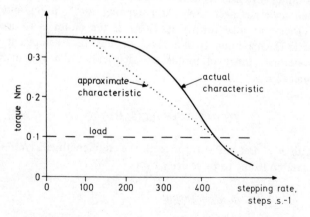

Fig. 6.8 *Pull-out torque characteristic showing linear approximations*

kg m² and a torque of 0·1 Nm. Derive the velocity profile for optimum acceleration of this system.

The pull-out rate for this system is 435 steps s^{-1}, because at this speed the motor's pull-out torque is equal to the load torque.

The variation of pull-out torque with speed can be approximated by the straight lines shown dotted in Fig. 6.8, i.e. $T(f)$ is approximated by the functions:

$$T(f) = 0·35 \text{ Nm} \qquad\qquad 0 < f < 100 \text{ steps s}^{-1}$$

and $\qquad = 0·426 - 0·000\ 75 \times f \text{ Nm} \qquad 100 < f < 435 \text{ steps s}^{-1}$

$$T_L(f) = 0·1 \text{ Nm}$$

For stepping rates up to 100 steps s^{-1}:

$$A = \int_0^f \frac{df}{T(f) - T_L(f)} = \int_0^f \frac{df}{0·25} = 4f \text{ (Nm s)}^{-1}$$

The step length is 15 degrees = 0·262 radians = $2\pi/np$. From Eqn. (6.9), therefore, the time taken to reach a stepping rate f is:

$$t = 2\pi JA/np = 0·262 \times 2 \times 10^{-4} \times 4f \text{ s} = 0·21f \text{ ms}$$

So the speed increases linearly with time up to 100 steps s^{-1}, which is attained after 21 ms.

For stepping rates between 100 and 435 steps s^{-1}:

$$A = \int_0^f \frac{df}{T(f) - T_L(f)} = \int_0^{100} \frac{df}{0·25} + \int_{100}^f \frac{df}{0·426 - 0·000\ 75f}$$

$$= 9630 - 1333 \log(1305 - 3f) \text{ (Nm s)}^{-1}$$

From Eqn. (6.9) the time taken to reach a stepping rate in the range 100–435 steps s^{-1} is therefore:

$$t = 2\pi JA/np = 0·0262 \times 2 \times 10^{-4} \times [9630 - 1333 \log(1305 - 3f)] \text{ s}$$

$$= 0·503 - 0·0697 \log(1305 - 3f) \text{ s}$$

The velocity profile can be constructed by evaluating this expression for some sample values of f:

f (steps s^{-1})	t (ms)
100	21
150	32
200	46
250	61
300	84
350	117
400	177

The complete profile for acceleration from rest is shown in Fig. 6.9. As these

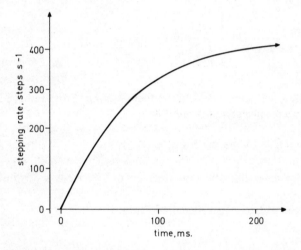

Fig. 6.9   *Velocity profile for acceleration from rest*

results are derived from the approximate pull-out torque characteristic which assumes a torque rather less than the actual pull-out torque, the motor should be able to follow this velocity profile even if the load is slightly more than expected.

When the system is decelerating the motor must produce a negative torque and therefore each phase must be switched on after the rotor has passed the phase equilibrium position. The load angle defined in Chapter 5 is then negative and the motor produces a braking torque $T_B(f)$, which is generally greater than the pull-out torque $[T_B(f) > T(f)]$. The equation of motion for the system becomes:

$$-T_B(f) = T_L(f) + J(d^2\theta/dt^2) \tag{6.10}$$

where $T_B(f)$ is the decelerating torque at stepping rate f. Substituting for $d\theta/dt$ in terms of f from Eqn. (6.7) and re-arranging:

$$df/dt = -[T_B(f) + T_L(f)] \times np/(2\pi J)$$

So the load torque assists the motor torque in decelerating the system inertia. The velocity profile during deceleration can be obtained by integrating the above equation:

$$\frac{np}{2\pi J} \int_0^t dt = \frac{npt}{2\pi J} = -\int_{f_I}^f \frac{df}{T_B(f) + T_L(f)} \qquad (6.11)$$

where t is the time taken to decelerate to a stepping rate f from an initial stepping rate f_I. Graphical methods must again be used to integrate the function $1/[T_B(f) + T_L(f)]$.

In some open-loop control schemes the precise times of phase excitation changes are required and the velocity profile must then be integrated to give the position/ time response of the system. A typical velocity profile and its integral are shown in Fig. 6.6, in which the load is accelerated to the pull-out rate and runs at this velocity until near the target position. Deceleration is initiated at a position which gives the motor sufficient time to decelerate the load inertia, so that the target position is attained with a velocity below the start-stop rate.

The sudden transition from operation at the pull-out rate, where the motor is producing the positive pull-out torque, to maximum deceleration, where the motor is producing the negative pull-out torque, is achieved by 'jumping' back the excitation sequence half of a complete cycle. This procedure can be understood most easily by referring to the static torque/position characteristics, even though at the speeds in question these characteristics are not strictly applicable. Fig. 6.10

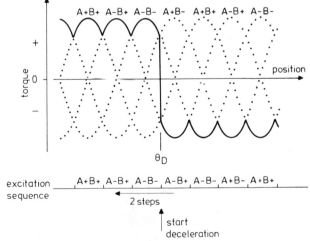

Fig. 6.10 *Excitation sequence to initiate deceleration*

shows the static torque characteristic for a hybrid motor excited two-phases-on. Initially the switching between phases occurs at the rotor positions corresponding to positive 'crossovers' of the characteristics, so the motor is developing the pull-out torque and the excitation sequence is $A+B+$, $A-B+$, $A-B-$, $A+B-$, $A+B+$, ... At the position θ_D, however, the motor has to produce its maximum negative torque so that the system can decelerate. Instead of switching $A-B-$ to $A+B-$ the next phases to be excited are $A-B+$, which enables the motor to develop maximum negative torque at that position. In this case the excitation sequence has been jumped back two steps (= half the total sequence of four steps) and the step counter contents must be adjusted.

Alternatively deceleration can be initiated by extending a phase excitation time until the rotor moves forward of the equilibrium position and the motor produces negative torque. In Fig. 6.10 deceleration can be initiated by prolonging the excitation time of phases $A-B-$ for approximately two step intervals. At the end of this time the rotor has moved forward to a position where the motor is producing negative torque. The transition to deceleration is slower than for the excitation jump method, because the motor torque only changes sign in response to a rotor movement, rather than an excitation change; but the complication of decrementing the step count is avoided.

6.4   Implementation of open-loop control

Four common methods of generating the open-loop velocity profile are examined in this Section. The choice of method for any particular system is a complex decision, in which the conflicting demands of high performance and low cost are evident. For example, an exponential ramp may be required for optimum acceleration, but its implementation is expensive and so the designer may compromise with a linear ramp, which is available at very low cost. Another factor is the rapid development of integrated circuit technology, which has made available a wide range of circuit functions at attractive prices, so that it is even realistic to assess the potential advantages of microprocessor-based control. Whilst present cost considerations may exclude the use of a dedicated microprocessor it may be possible to direct any spare processing capability towards stepping motor control. In this latter case the incremental costs of extra memory and software development is likely to be less than the price of a separate hardware controller.

6.4.1   Microprocessor generated timing

The microprocessor is well-suited to the generation and timing of the digital signals required for stepping motor control and in Chapter 8 the whole subject of microprocessor/stepping motor interfacing is discussed in detail.

As far as open-loop control is concerned, however, the use of a dedicated microprocessor for a single motor would be unnecessarily expensive. With open-loop control even lightly-loaded motors can rarely operate at speeds of more than 10 000 steps per second and therefore the microprocessor need only issue a step command every 0·1 ms. As the program time needed to send each step command is likely to be much less than 0·1 ms the processor has spare capacity for other tasks. Efficient utilisation of processing capacity is obtained by using an interrupt routine to control the motor, with the main program being interrupted by a constant frequency clock set at a suitable multiple of the pull-out rate.

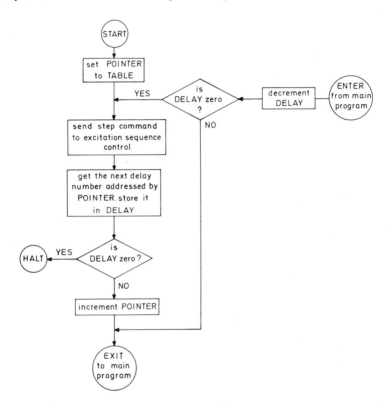

Fig. 6.11   *Flowchart for microprocessor-based open-loop control of a stepping motor*
Sample table of DELAY values for a movement of 14 steps

30	10
23	10
18	25
14	13
12	19
11	27
10	0

A stepping motor control program flowchart is shown in Fig. 6.11. In this example the number of steps executed is fixed and the times between step commands are controlled by the values stored in the look-up table beginning at location TABLE. The program starts by setting a register POINTER equal to TABLE, so the register contains the location of the first value in the table. The first step command is then sent to the excitation sequence control, which changes the phase excitation in the motor.

There must now be a delay before the next step command is issued, so that the motor has time to execute the first step. The first value is collected from the look-up table and stored in location DELAY, which is checked to ensure it is not zero, because a zero value indicates that the end of the table has been reached. The register POINTER is incremented and therefore points to the next table value. Control is then returned to the main program.

Execution of the main program continues until the next clock interrupt, which returns the processor to the motor control routine at ENTER. The value of DELAY is decremented and compared to zero. If DELAY is not zero control is immediately returned to the main program, but if DELAY has reached zero the next step command is sent to the excitation sequence control and DELAY is loaded with the next value from the look-up table. The times between step commands are therefore proportional to the (constant) clock period and the look-up table values. For example, after the first step command DELAY is loaded with 30 and therefore 30 clock interrupts occur before DELAY is counted down to zero.

The look-up table values are chosen to ramp up the motor velocity over six steps to a maximum stepping rate which is 1/10 of the clock frequency. Deceleration commences with a long delay number (25), which allows time for the rotor to swing past the equilibrium position into a position where the motor is producing the negative torque required for deceleration. Only four steps are required to decelerate the motor, as the load torque contributes to the decelerating torque. The system finally comes to rest fourteen steps from the initial position, with the program detecting a zero value of DELAY and exiting to HALT.

In this example the distance travelled is fixed, but the program could allow the target position to be loaded before the travel commenced. Additional steps can be produced by expanding the look-up table to include more high-speed (short-delay) values. However if the target is less than fourteen steps from the original position the appropriate number of look-up table entries can be deleted, starting with the short-delay values.

The look-up table entries can be calculated, using the methods described in Section 6.3 to determine the optimum velocity profile. Alternatively the table values can be found experimentally and, as an aid to this potentially laborious process, Lafreniere (1979) has developed a microprocessor-based interactive system, which calculates the complete look-up table from a limited number of velocity profile parameters, such as maximum stepping rate, acceleration/deceleration rate.

Several motors can be controlled from the same microprocessor provided the execution time for each motor control routine is sufficiently short, as illustrated by the following example.

Example

A stepping motor system has a pull-out rate of 500 steps per second and a micro-processor is to be used for its open-loop control. How is the clock interrupt rate determined? If the step command routine can be executed in 30 μs, how many motors could be controlled by the microprocessor?

Suppose the delay number stored in the look-up table is m for the maximum stepping rate of 500 steps per second. The clock rate is then $500m$ Hz, because there are m clock cycles per step command. The next lowest stepping rate corresponds to a delay number of $m+1$ and is therefore $500m/(m+1)$ steps per second. The motor must be able to respond to the instantaneous change in stepping rate, as a result of a change in delay number from $m+1$ to m; the transient performance of the system following this change is best investigated experimentally. If, for example, the system is just able to respond to a change in stepping rate from 475 to 500 steps per second:

$$500m/(m+1) = 475$$

and so $m=19$. The clock interrupt rate must be set at $500m$ Hz = 9·5 kHz.

At a clock frequency of 9·5 kHz the interrupts occur at intervals of approximately 105 μs. The number of motors which can be controlled is limited by the requirement that step commands may have to be issued to all motors between successive clock interrupts. Each step command is issued in 30 μs so the step commands to 3 motors are issued in 3 x 30 = 90 μs, but four motors would require 4 x 30 = 120 μs. Therefore no more than three motors can be controlled by a single microprocessor.

6.4.2 Hardware timing

If the acceleration of a system occurs over a small number of steps then the phase excitation timings can be generated by digital integrated circuits. In Fig. 6.12(a), for example, a sequence of variable-duration delays gives precise timing of the first three steps, which are used to accelerate the motor to a stepping rate defined by the system clock frequency. A further sequence of delays is used to decelerate the motor as it approaches the target position.

With the system initially at rest a pulse is applied to the START input. This pulse is applied directly to the phase sequence generator, via a series of logical OR gates, and the consequent excitation change initiates acceleration of the motor. The starting pulse also triggers the first delay circuit, which delays the pulse for a time T_1, during which the motor moves to the first phase switching position. The pulse output of the first delay is fed to the phase sequence generator and also triggers the next delay circuit. This sequence continues until all the delays have operated. The output of the last delay is used to start the constant frequency clock, which produces the subsequent step commands, as shown in Fig. 6.12(b).

At the beginning of the operation the target position is loaded into the downcounter. Each step command, from either the delay circuits or the clock, decrements the downcounter, which therefore records the number of step commands to be

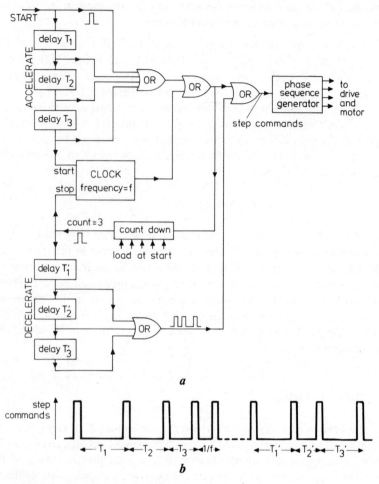

Fig. 6.12 *Open-loop control by hardware timing*
   a   system block diagram
   b   timing of step commands

issued before the target is reached. When the number of steps to be executed is equal to the number of deceleration delay circuits the downcounter produces a pulse, which switches off the clock and also triggers the first deceleration delay, T_1'. Controlled deceleration over the final steps to the target is provided by the three variable-duration delays T_1', T_2', T_3', which are triggered in sequence and produce step commands for the phase sequence generator. Deceleration commences with a long excitation period (T_1') to allow the rotor to move ahead of the equilibrium position and produce negative torque.

There is, of course, no reason why the number of delays should be limited to the three illustrated in Fig. 6.12. If the maximum operating speed of the system is to approach the pull-out rate then between 20 and 50 delays might be required and the techniques would not be cost-effective. In general the use of hardware timing is restricted to applications requiring a modest increase in the operating speed above the normal starting/stopping rate. Under these circumstances the delay times can be successfully predicted from static torque/rotor position characteristics (Lawrenson *et al.* 1977).

6.4.3 Pulse deletion

This acceleration/deceleration scheme does not allow the fine control of excitation timings available with the other methods, but it does have the merit of simplicity and consequent low cost. The system block diagram is shown in Fig. 6.13(*a*) and

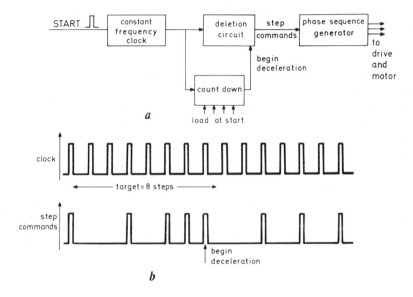

Fig. 6.13 *Open-loop control by pulse deletion*
a system block diagram
b clock signal and step commands

the associated timing diagram in Fig. 6.13(*b*). At the centre of this system is a constant frequency clock set at a rate which is substantially above the starting/ stopping rate of the system. The clock pulses are sent to a downcounter which is, as usual, loaded with the target position before the operation starts. The clock pulses are also input to the pulse deletion circuit, which blocks some of the clock signals during acceleration and deceleration.

In the example shown in Fig. 6.13 a total of three clock pulses are blocked during the acceleration interval. The first clock pulse is passed directly to the phase sequence generator, causing the change in excitation which starts acceleration. A simple clocked flip-flop circuit (Maginot and Oliver, 1974) blocks the next two clock pulses, so the next change in excitation is caused by the fourth clock pulse. The time available for the motor to travel to the appropriate phase switching position is therefore three times the clock period. The fifth clock period is also blocked, so there is a delay of two clock periods between the second and third step commands. All subsequent clock signals are transmitted and the motor then operates at a constant stepping rate equal to the clock rate. At this stage, with acceleration complete, three clock pulses have been deleted, but all the pulses have been sent to the downcounter, which therefore records a position three steps nearer to the target than the actual motor position.

When the counter reaches zero the 'missing' three step commands are used to decelerate the motor to the target. The clock pulses are still input to the deletion circuit and therefore the process of deleting pulses can be repeated. A long delay is required at the beginning of deceleration so that the rotor can move ahead of the appropriate equilibrium position and produce negative torque. This long delay is obtained by deleting two clock pulses. Subsequent deceleration can be faster than acceleration, because friction torques assist retardation, and remaining clock pulses are therefore issued at half the clock frequency.

Faster operating speeds can be obtained with a higher clock frequency and more stages of pulse deletion. However the speed range is ultimately restricted (as with microprocessor-based control) by the discrete speeds available from a single clock and the ability of the system to accelerate between these speeds within one step length.

Beling (1978) describes an alternative implementation, in which a clock signal — set at the maximum stepping rate — is input to a phase sensitive detector, which also receives step command signals. The detector output voltage is proportional to the difference between the two input frequencies and is fed to a voltage-controlled oscillator via an analogue low-pass filter. The frequency of step commands produced by the oscillator is ramped up to the clock frequency at a rate which is determined by the filter parameters and which can be matched to the motor/lead combination.

6.4.4   Analogue ramp up/down

The final technique for producing a velocity profile involves a voltage-controlled oscillator with the controlling voltage generated by an analogue circuit. In the circuit of Fig. 6.14(*a*), for example, linear accleration and deceleration profiles are produced by integrating a signal which turns on when the motor is to accelerate and turns off when deceleration in required. As long as the pull-out torque of the motor is constant with speed, the rate of acceleration is constant (as in example of Section 6.3) and a linear ramp produces optimum performance from the system.

With the motor initially at rest the target position is loaded into the downcounter and a START signal is applied to the input AND gate. The integrator input therefore shifts from a LOW to HIGH state, as shown in Fig. 6.14(*b*), and the integration commences. The integrator time constant can be varied by adjusting the timing

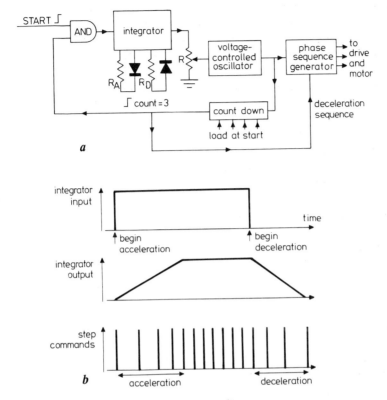

Fig. 6.14   *Open-loop control by analogue ramp up/down*
a   system block diagram
b   timing of integrator input/output and step commands

resistor R_A to give the required rate of acceleration. A linear ramp with the appropriate slope is input to the voltage-controlled oscillator, which generates the step commands at a linearly increasing rate. A variable attenuator (R) can be used to adjust the maximum voltage input to the oscillator and therefore controls the maximum stepping rate.

The step commands are also input to the downcounter, which records the instantaneous position of the system relative to the target. At a preset number of steps from the target the downcounter changes the state of the DECELERATE signal from HIGH to LOW. The AND gate output therefore becomes LOW and the integrator ramps down to zero. Rates of deceleration and acceleration can differ because independent timing resistors (R_D and R_A) are provided in the integrator.

One additional problem is that the voltage-controlled oscillator is unable to provide the pause between step commands needed at the beginning of the deceleration period so that the rotor can advance beyond the excited phase equilibrium position; the control of individual phase excitation timings available with the other open-loop schemes cannot be produced with the voltage-controlled oscillator technique. Therefore the DECELERATE signal is applied to the phase sequence generator, which shifts the phase sequence back by an appropriate number of steps. The rotor then finds itself in advance of the instantaneous equilibrium position and the motor is able to produce the negative decelerating torque.

Several adjustments are necessary when the system load is changed. Differences in load torque lead to a change of pull-out rate and the attenuator R must be set so that the maximum stepping rate does not exceed the new pull-out rate. The rates of acceleration and deceleration depend on the load torque and inertia and are controlled by the timing resistors R_A and R_D, which must be set accordingly. Finally the number of steps needed to decelerate from the maximum stepping rate to zero must be estimated and the downcounter set to generate the DECELERATE signal when the system is that distance from the target.

A linear ramp velocity profile is only optimal if the pull-out torque of the motor is constant up to the maximum required operating speed. In many systems the optimum profile is more involved, e.g. an exponential form may be needed for acceleration and an inverse exponential for deceleration. These profiles can be approximated by diode-shaping circuits (Maginot and Oliver, 1974) which 'round-off' the linear ramp up/down signal before it is input to the voltage-controlled oscillator.

6.5   Improving acceleration/deceleration capability

The rated current of a stepping motor winding is assessed on the basis of the maximum allowable temperature rise when the winding is continuously excited. This situation only arises when the motor is stationary; if the motor is moving the phases are excited in sequence and any one winding is excited for only a fraction of the cycle. Therefore the winding currents can be increased above the rated value when the motor is rotating, because the extra heat generated while the phase is turned on can be dissipated later in the cycle when the phase is off. Although a stepping motor is magnetically saturated at the rated current, larger currents can still improve the motor torque and the corresponding rates of acceleration and deceleration.

How can the supply voltage — and consequently the phase current — be increased during acceleration or deceleration? If the system incorporates a bilevel or chopper drive circuit the current control can be over-ridden at the appropriate times to make the full supply voltage available continuously. For the simpler forms of drive, however, the cost of additional supply capacity may not be justifiable in terms of the limited increase in performance. Under these circumstances the drive modification illustrated in Fig. 6.15 may be useful.

The circuit operates by storing energy in the capacitor C, which can be discharged rapidly into the phase windings when additional torque is needed. During normal operation the transistor switch S of Fig. 6.15 is open. The phase circuits are excited by the supply voltage V_{dc}, which produces the rated current in an excited phase when the motor is stationary, and the phase currents flow via the

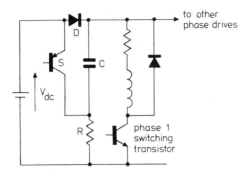

to other
phase drives

phase 1
switching
transistor

Fig. 6.15 *Circuit for 'voltage-boosting' during acceleration/deceleration*

diode *D*, which is forward-biased by the supply voltage. A small additional load arises from the charging of capacitor *C* at a rate limited by the resistance *R*.

If the supply voltage is to be augmented at any time the transistor switch *S* is closed. The base drive for the transistor can be derived from the START or DECELERATE control signals if fast acceleration/deceleration is required. The phase currents now flow from the supply through transistor *S* and capacitor *C*, so the effective phase voltage is the sum of the supply and capacitor voltages. Initially the capacitor voltage is V_{dc} and therefore the phase voltage is $2V_{dc}$, but this decays to V_{dc} as the capacitor discharges. The diode *D* is reverse-biased by the capacitor voltage so it can resume conduction of the phase currents as soon as the capacitor has discharged. When acceleration or deceleration is complete the transistor switch is opened and the capacitor can then recharge in preparation for the next speed change.

The values of the energy-storing capacitor (*C*) and current-limiting resistor (*R*) must be chosen carefully. Firstly, the size of capacitor is dictated by the phase current and the time for which 'voltage-boosting' is needed; large phase currents and long boost times lead to large capacitor values. The resistance *R* limits the additional current drawn from the supply to charge the capacitor, when the transistor switch is initially opened. If the capacitor is completely discharged the initial current is V_{dc}/R, so a high resistance reduces the charging current. However the time taken to re-charge the capacitor is proportional to the charging time constant (*RC*), so if a high resistance is used the capacitor may not be completely charged when its stored energy is next required. Therefore the choice of *R* must be a compromise which limits the charging current to within the supply capacity, but ensures that the capacitor is fully charged as quickly as possible.

The 'voltage-boosting' scheme can be used in association with any type of control (including closed-loop) and typically produces a 50% improvement in acceleration/deceleration capability (Lawrenson *et al.* 1977).

Example

A three-phase stepping motor is excited one-phase-on by its rated current of 2 A.

The d.c. power supply is 20 V, 2 A. Find the size of capacitor required to produce voltages of 30-40 V during acceleration/deceleration periods of 10 ms at 360 ms intervals. What would be the new current capacity of the d.c. power supply?

During speed changes the supply voltage may decrease from 40 V to 30 V and therefore the phase current reduces from 4 A to 3 A. For simplicity assume a linear voltage decay, so the average current (I) is 3·5 A and the rate of change of voltage with time is $(40 - 30)/10^{-2}\,\mathrm{V\,s^{-1}} = 1\cdot0\,\mathrm{kV\,s^{-1}}$ if the capacitor value is C:

$$I = C\,dV/dt$$

$$C = I/(dV/dt) = 3\cdot5/(1\cdot0 \times 10^3)\,\mathrm{F} = 3500\,\mu\mathrm{F}$$

The periods of constant speed operation are 360–10 ms = 350 ms in duration and about 5 time constants ($5RC$) are required for the capacitor to re-charge:

$$5RC = 350\ \mathrm{ms}$$

$$R = 350 \times 10^{-3}/(5 \times 3\cdot5 \times 10^{-3})\ \mathrm{ohms} = 20\ \mathrm{ohms}$$

The maximum charging current occurs immediately after the transistor switch is opened. The d.c. supply voltage (20 V) is applied to the series combination of the resistor R and capacitor C, which has a residual charge of 10 V:

$$\text{Maximum charging current} = (V_{dc} - V_C)/R = (20 - 10)/20\ \mathrm{A}$$

$$= 0\cdot5\ \mathrm{A}$$

Therefore the current capacity of the supply must be increased from 2·0 A to 2·5A.

Chapter 7

Closed-loop control

7.1 Introduction

In a closed-loop stepping motor system the instantaneous rotor position is detected and fed back to the control unit. Each step command is issued only when the motor has responded satisfactorily to the previous command and so there is no possibility of the motor losing synchronism.

A schematic closed-loop control is shown in Fig. 7.1. Initially the system is

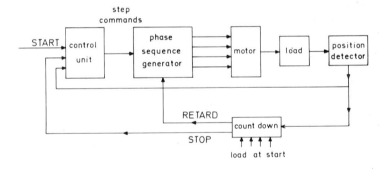

Fig. 7.1 *Closed-loop control of a stepping motor*

stationary with one or more phases excited. The target position is loaded into the downcounter and a pulsed START signal is applied to the control unit which immediately passes a step command to the phase sequence generator. Consequently there is a change in excitation and the motor starts to accelerate at a rate dictated by the load parameters.

As the first step nears completion the position detector generates a pulse which is sent to both the downcounter and the control unit. The downcounter is decremented and therefore contains the position of the load relative to the target. Note here the contrast between closed- and open-loop control schemes; with open-loop control the downcounter is able to record only the number of step commands sent to the motor and there is no guarantee that these steps have been executed.

With closed-loop control, however, the downcounter is recording actual load position.

The position detector pulse sent to the control unit is used to generate the next step command. For larger loads the time taken to reach the first step position is longer and therefore the time between successive step commands is automatically adjusted to allow for the slower rate of acceleration. The motor eventually reaches a maximum operating speed, which, as in the open-loop case, is dictated by the motor and load torque/speed characteristics, and continues to run at this speed until the target position is approached.

The downcounter is used to initiate the change in phase sequence required to decelerate the motor to the target. The number of steps executed during deceleration depends on the load conditions — a load with high inertia and low torque needing more steps to decelerate than a low inertia, high torque load — but generally the worst load conditions are anticipated in setting the count required to generate the RETARD signal. If the actual load conditions are less severe than this worst case the system can be run at a stepping rate below its starting/stopping rate until the target position is attained. When the downcounter reads zero the required number of steps have been executed and a STOP signal is sent to the control unit to inhibit all further step commands.

A closed-loop control therefore matches the excitation timing to the load conditions and is capable of giving a near optimal velocity profile with consequent rapid load positioning (Kuo, 1974). To many users its most attractive feature is that load position is monitored directly, so that even under the worst load conditions there is no possibility of losing synchronism between step commands and rotor position.

The effective implementation of closed-loop control was inhibited, until recently, by a number of problems, which are discussed in the subsequent sections of this Chapter. The position detector produces a pulse for every step executed by the system, but if the motor torque is to be maximised the phase of the position pulse must be related to the instantaneous operating speed. At high speeds, for example, the pulses must occur earlier in the cycle if the phase current is to attain its rated value before the rotor position for maximum phase torque is reached.

In any closed-loop system the position detector represents a considerable proportion of the total cost. Traditionally position detection is achieved by optical transducers (Lajoie, 1973), but the high costs of this method, particularly for small step angle motors, have prompted the investigation of several alternatives. Among the most attractive of the alternatives is 'waveform detection', in which modulation of the phase currents by the motional voltage is used to detect rotor position (Frus and Kuo, 1976; Jufer, 1976; Bakhuizen, 1979).

7.2   Switching angle

7.2.1   Switching angle to maximise pull-out torque

In a closed-loop system a position detector generates a pulse which signals to the control unit that a step has been completed. The question then arises of exactly which position should be detected?

At low operating speeds the optimum detected position can be deduced from the static torque/rotor position characteristic for the excitation scheme being employed. For example, one-phase-on torque/position characteristics for a three-phase, 15-degree step motor are illustrated in Fig. 7.2, which shows that torque

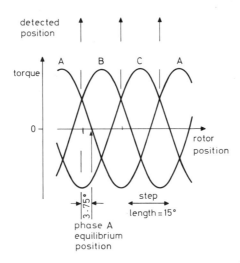

Fig. 7.2 *Detected position setting to maximise torque at low speeds*

developed by the motor can be maximised by arranging for position pulses to be generated at the crossover points of the phase torque characteristics.

If the motor is to operate at higher speeds the static torque/rotor position characteristics are no longer a reliable guide, because they do not account for distortion of the current waveform due to the winding time constant and motional voltage. In these circumstances each change in phase excitation must occur earlier relative to the rotor position, so that the phase current has sufficient time to become established before the rotor reaches the position of maximum phase torque. This process is commonly referred to as 'ignition advance'. The relationship between pull-out torque and operating speed has been examined in Chapter 5 and some of the results from that analysis can be applied to the closed-loop control problem.

For both hybrid and variable-reluctance stepping motors the fundamental component of phase voltage is [from Eqns. (5.7) and (5.24)] :

$$v = V \cos \omega t \tag{7.1}$$

and the time variation of rotor position is [from Eqns. (5.5) and (5.20)] :

$$p\theta = \omega t - \delta \tag{7.2}$$

where ω is the angular frequency of the supply fundamental component, p is the number of rotor teeth and δ is the load angle. For maximum (pull-out) torque at a given speed, the load angle is related to the average winding inductance (L) and the total phase resistance (R) by Eqn. (5.13):

$$\delta = \tan^{-1}\omega L/R \tag{7.3}$$

At low speed ω approaches zero so, from Eqn. (7.3), $\delta=0$ and the appropriate rotor position for the detected pulses can be predicted from the torque/position characteristics. For higher speeds the load angle d increases until at the highest speeds it approaches 90 degrees. Comparing Eqns. (7.1) and (7.2) we see that the phase relationship between the fundamental component of phase voltage and the time variation of rotor position must be equal to δ/p if the pull-out torque is required. This phase relationship is determined by the point at which the position detector generates a pulse to trigger a change in excitation and so the detector trigger point must vary with speed if the pull-out torque is required over the complete speed range. It is convenient to introduce the idea of 'switching angle', which is simply the *change* in detected rotor position relative to the low-speed detected position. The pull-out torque is maximised at a supply angular frequency ω when the switching angle is set to its optimum value:

$$\text{Optimum switching angle} = (\tan^{-1}\omega L/R)/p \tag{7.4}$$

The variation of this switching angle with supply frequency is illustrated in Fig. 7.3.

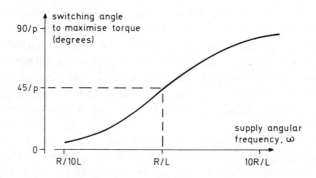

Fig. 7.3   *Optimum switching angle vs. supply angular frequency*

If the switching angle is not optimised at any operating speed then the motor torque is less than the pull-out torque. The torque reduction can be predicted from the results of Chapter 5, as the effective load angle is simply p times the switching angle. For example Eqn. (5.13) gives the torque produced by a hybrid motor when operating at any supply frequency and load angle. The variation of torque with switching angle and speed for a three-phase, 15-degree step motor ($p=8$) is shown in Fig. 7.4. These characteristics include a negative value of switching angle, for which

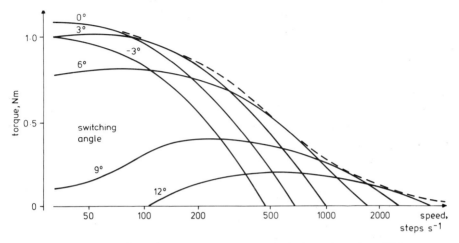

Fig. 7.4 *Typical torque/speed/switching angle characteristics for a three-phase 15 degree step-angle motor with one-phase-on excitation*
— — — pull-out torque/speed characteristic

the phase excitation is changed after the static torque/position 'crossover' points. Fig. 7.4 shows quite clearly that an injudicious choice of switching angle may prevent the motor from producing any torque at certain speeds, e.g. switching angle = 12 degrees at speeds below 100 steps per second. Although large switching angles must be used to attain high pull-out rates, these angles are inappropriate for low-speed operation.

The need for a speed-dependent switching angle is a major obstacle to the successful implementation of closed-loop control systems. Most position detectors produce a pulse at a fixed position, which must be chosen according to the application and type of control required. The factors relevant to the choice of switching angle are discussed in the following Sections.

7.2.2 Position control

If a closed-loop stepping motor system is to position a load in the shortest possible time it must accelerate rapidly to a high speed, but the choice of switching angle is subject to a conflict between the demands of fast acceleration and high speed operation. For small switching angles a high torque is developed at low speeds, so the system accelerates rapidly from rest, but the maximum speed is restricted. In Fig. 7.4, for example, a switching angle of 3° gives a torque of about 1 Nm at low speeds, but the motor cannot operate above 1000 steps per second.

Conversely if a large switching angle is chosen the torque is small at low speeds and the initial rate of acceleration is slow, but ultimately higher running speeds can be attained. The characteristics of Fig. 7.4 show that for a switching angle of 9° the low speed torque is only 0·1 Nm, but speeds of up to 2500 steps per second can be reached if the motor is lightly loaded.

For fast positioning, therefore, the optimum performance from a closed-loop system can only be obtained if an ignition advance scheme is used to increase the switching angle with speed. With the presently available methods of position detection – described in Section 7.3 – continuous ignition advance is impractical. The detector is generally limited to generating one pulse per step at a fixed point relative to the phase equilibrium position, although in more sophisticated systems it is possible to generate several position pulses per step and choose the switching angle best suited to the instantaneous speed.

The choice of a fixed switching angle depends on the motor/load parameters and the distance to be travelled. If the target position is relatively few steps from the initial position or the load inertia is high, the system is unable to accelerate to a high speed. The most important consideration is that a high torque should be available at low speeds and therefore a small switching angle is chosen. This argument is reversed when the load has to move a large distance, because a high operating speed is then required. The time taken to reach the highest speed is small compared to the time spent operating at this speed, so a large switching angle is chosen. The consequent reduction in torque at low speeds, resulting in poor initial acceleration, is compensated by the higher steady-state speed.

Example

The motor with characteristics shown in Fig. 7.4 is to drive a load of 0·5 Nm and inertia 10^{-4} kg m^2 a distance of 60 steps under closed-loop control. Estimate the fixed switching angle which minimises the time taken to reach the target.

Referring to Fig. 7.4, the pull-out rate for the motor with a load torque of 0·5 Nm is approximately 540 steps per second. At this speed the optimum switching angle is 6 degrees and therefore the choice of switching angle is limited to the range 0–6 degrees. The characteristics for angles of 0, 3 and 6 degrees are reproduced in Fig. 7.5.

The maximum speed attainable with each switching angle is simply the stepping rate at which the appropriate characteristic intersects the 'load torque' line:

Switching angle	Maximum speed	
(degrees)	(steps s^{-1})	(rads s^{-1})
0	316	82·6
3	436	114·0
6	436	141·5

The velocity profile during acceleration can be calculated precisely for each switch-

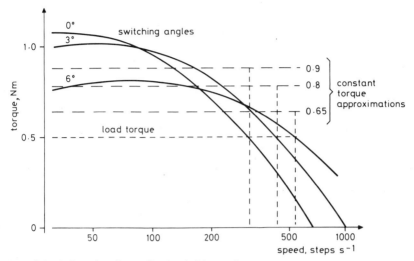

Fig. 7.5   *Calculation of optimum fixed switching angle*

ing angle using the graphical method described in Section 6.3. However a rough estimate of optimum switching angle can be obtained if the three torque/speed characteristics are approximated by constant torques, effective up to the maximum speed for each angle:

Switching angle (degrees)	Torque (Nm)
0	0·90
3	0·80
6	0·65

The time taken and number of steps executed during acceleration to the maximum speed may be calculated, since for a constant torque, T:

$$T - T_L = J \, d\omega/dt$$

$$T - 0.5 = 10^{-4} \, d\omega/dt$$

$$t = \int_0^\omega \frac{10^{-4}}{(T-0.5)} \, d\omega = \frac{10^{-4}\omega}{(T-0.5)}$$

For constant accelerating torque the instantaneous velocity is proportional to time and therefore the distance travelled during acceleration is simply:

Distance = maximum speed x time/2

so for each switching angle the acceleration time and distance can be found:

Switching angle (degrees)	Time taken to reach max. speed (ms)	Distance travelled during acceleration (steps)
0	20·6	3
3	38·0	8
6	94·3	25

The time taken to travel a total distance of 60 steps can be calculated for each switching angle, e.g. if the switching angle is 6 degrees the system executes the first 25 steps in 94·3 ms and the remaining 60 – 25 = 35 steps at a constant speed of 540 steps per second:

Switching angle (degrees)	Acceleration time (ms)	Distance travelled at max. speed (steps)	Time at max. speed (ms)	Total time (ms)
0	20·6	57	180	200
3	38·0	52	119	157
6	94·3	35	65	159

The time taken to decelerate has been neglected in this calculation because the retarding torque (motor braking torque + load torque) is, in all three cases, much larger than the accelerating torque (motor accelerating torque – load torque).

Switching angles of 3 and 6 degrees produce a faster travel than an angle of 0 degrees. More detailed calculations (taking into account the variation of torque with speed) for angles of 3 and 6 degrees would reveal the optimum setting for the switching angle.

7.3   Detection of rotor position

7.3.1   Optical detection

The most popular type of position detector is the incremental optical encoder, which is shown schematically in Fig. 7.6. In the basic device an opaque disc is fastened to the shaft and therefore rotates at the same speed as the motor. Around the edge of the disc are radial slots, which are equal in number to the steps per revolution of the motor. A light source and photosensitive device — photodiode, phototransistor or photovoltaic cell — are placed on opposite sides of the disc, so that the device is illuminated whenever a disc slot is in front of the light source. The photodevice therefore generates a signal once per step movement of the rotor.

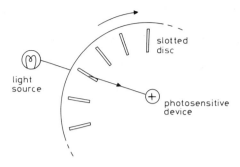

Fig. 7.6 *Schematic section of an incremental optical encoder*

Incremental encoders of this type are usually supplied as sealed units so that dirt cannot interfere with the operation of the optical system and the light source/detector alignment cannot be disturbed. The required detected position is obtained by ensuring that the slotted disc is connected to the motor shaft with the correct relative orientation.

The signal produced by the photosensitive device depends on the level of illumination, which varies approximately linearly as a disc slot moves into alignment with the light source (Fig. 7.7). Therefore the photodetector generates a triangular pulse, which must be processed by a comparator to give a sharp rectangular pulse

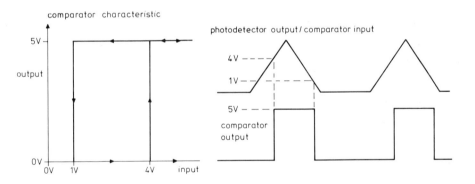

Fig. 7.7 *Processing the position detector signal*

suitable for triggering the control unit. A comparator with user-adjustable switching level is often incorporated in the encoder unit and small variations in the detected position can be obtained by adjusting this switching level. If the motor oscillates around the detected position when coming to rest a sequence of pulses would be generated, leading to a false count of steps executed. This hazard is avoided by introducing hysteresis into the comparator characteristics — as shown in Fig. 7.7 — so that small changes in the signal from the photosensitive device do not give further detected pulses.

One development of the optical encoder employs two photosensitive devices with a small relative displacement. The pulse-shaped output from the detectors have a slight phase displacement, the sign of which depends on the direction of rotation. In Fig. 7.8(*a*) signal X leads signal Y when the motor is moving in the

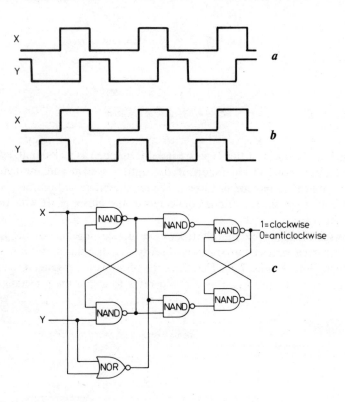

Fig. 7.8   *Direction of rotation deduced from position detector signals*
 a   clockwise rotation
 b   anticlockwise rotation
 c   derivation of direction signal

clockwise direction, but in Fig 7.8(*b*) Y leads X for anticlockwise rotation. The logic circuit illustrated in Fig. 7.8(*c*) processes the two signals and generates a DIRECTION signal which indicates the instantaneous direction of rotation. This signal can be compared to the signals being produced by the phase sequence generator as a check that the motor is operating correctly.

The cost of incremental encoders depends on the number of slots on the rotating disc. For motors with large step angles (e.g. 15 degrees, 24 steps per revolution) the number of slots is low and therefore incremental encoders are an attractive form of position detection. However in hybrid stepping motor systems the step angles are small (typically 1·8 degrees, 200 steps per revolution) and the price of the encoder can then be a significant proportion of the total closed-loop system costs.

Encoders generating up to 5000 pulses per revolution may be used in more sophisticated closed-loop systems where several pulses are generated for each step of the motor. The pulse used to trigger the control unit can be varied according to the time between successive pulses, giving control of switching angle with operating speed.

7.3.2 Waveform detection

The modulation of winding currents by the motional induced voltage is the basis of a scheme known as waveform detection. In Chapter 5 the voltage induced in the windings was shown to have an important influence on the current waveforms of both hybrid and variable-reluctance stepping motors. As the instantaneous induced voltage is a function of rotor position an analysis of the current waveform, with particular regard to the effects of induced voltage, reveals information about the instantaneous rotor position.

The block diagram of a closed-loop waveform detection system is shown in Fig. 7.9. The winding currents are monitored by measuring the voltage drops across small resistances connected in series with the windings. A waveform analyser

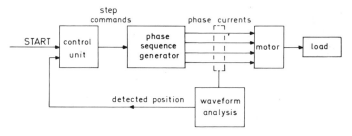

Fig. 7.9 *Closed-loop control by waveform detection*

processes the voltage signals and returns position detector pulses to the control unit at the required positions. Several advantages of waveform detection are apparent:

(*a*) No additional mechanical connections to the motor are required.

(*b*) The waveform analyser can be situated with the drive and control, which may be a considerable distance from the motor.

(*c*) The detector costs are substantially reduced in comparison to an incremental optical encoder, as the waveform analyser is composed of simple electronic circuitry.

The principles of waveform detection can be illustrated by referring to a three-phase variable-reluctance stepping motor with one-phase-on excitation. Typical high-speed current waveforms are shown in Fig. 7.10, in which the d.c. supply voltage is applied to phase A for 1/3 of the excitation cycle. Similar voltage waveforms apply to windings B and C with an appropriate phase displacement, so that winding C, for example, is turned on at the mid-point of the winding A freewheeling interval.

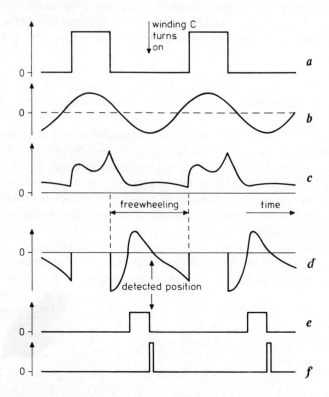

Fig. 7.10   *Waveform detection for phase A of a three-phase motor with one-phase-on excitation*
   a   applied phase voltage
   b   motional voltage
   c   current waveform
   d   rate of change of freewheeling current
   e   comparator output
   f   position detector pulses

   If the motor is operating as part of a position control system the pull-out torque must be produced at all speeds; any surplus of motor torque over load torque accelerates the system to a higher speed and thereby reduces the positioning time. For maximum torque the motional voltage must be at its maximum during the phase winding excitation interval [Fig. 7.10(*b*)]. A positive motional voltage opposes the flow of current in the winding, so at high speeds the winding current may even decay as the motional voltage reaches its maximum value. During the freewheeling interval the motional voltage becomes negative and the freewheeling winding current may be forced to reverse direction. As the motional voltage passes its maximum negative value the freewheeling current is able to continue its decay [Fig. 7.10(*c*)].
   The waveform analysis circuit is set to detect the zero rate of change of free-

wheeling current in each winding. This is accomplished by first differentiating the freewheeling current signal and then using a comparator to find the time instants at which there is no change of current. In Fig. 7.10(*d*) the rate of change of freewheeling current is large and negative when the winding is turned off, but the decay is reversed by the motional voltage and the rate of change reaches a positive maximum. Eventually the motional voltage reduces and the rate of change of current is again zero, causing a change in the comparator state. This transition in the comparator output is used to generate the position detector pulse [Fig. 7.10(*f*)], which occurs at the time when winding *C* should be switched on and therefore can be used directly as a step command. Similar position detector signals can be derived from the freewheeling currents of the other two windings, i.e. freewheeling currents in winding *B* generate a trigger signal for winding *A* and winding *C* currents produce step commands for winding *B*.

Although the principle of waveform detection has been discussed here with particular reference to a three-phase variable-reluctance motor, it should be apparent that the method can be applied to any stepping motor in which the motional voltage influences the current waveform. The setting up of a waveform detection system is usually based on experimental observation of the current waveform with open-loop control and pull-out load torque. A suitable feature of the waveform can be selected for generation of the position signal over the required speed range (Kuo and Cassat, 1977). In making this choice zero rates of change of current are to be preferred, as they occur at rotor positions which are almost independent of instantaneous stepping rate.

A number of difficulties can occur when implementing a waveform detection scheme. At low speeds the magnitude of the motional voltage may be insufficient to reverse the direction of freewheeling current and an alternative detection system may be required for this portion of the speed range. A similar problem occurs during deceleration when the phase relationship of the current and motional voltage is reversed, giving very different current waveforms. Frus and Kuo (1976) describe a system in which the motor coasts to rest with one winding continuously excited, giving position pulses suitable for recording distance travelled but not for triggering changes in excitation. There is considerable scope for innovation in this area and microprocessor-based waveform analysis could improve the flexibility of this position detection scheme.

Waveform detection based on freewheeling currents is inappropriate for the bilevel and chopper drives described in Section 5.4. In these circuits the freewheeling path includes a high supply voltage, so the freewheeling currents are forced to zero before the motional voltage reaches its negative maximum. With these high-performance drives the variation of current during the interval when the winding is excited must be used as the basis for a detection scheme. For a chopper drive the winding current variations are small, but when the motional voltage is near its positive maximum value the supply has to be connected to the winding for longer time intervals to maintain the rated current. Therefore waveform detection can be established from the variation of chopping frequency over the winding excitation interval.

Chapter 8

Microprocessor-based
stepping motor systems

8.1 Introduction

Stepping motors are often used as output devices for microprocessor-based control systems, e.g. a graphplotter pen may be driven by x- and y-axis stepping motors which are controlled by a microprocessor. The essential feature of these systems is that the microprocessor program produces a 'result' and the stepping motor must then move the load to the position corresponding to this 'result'. In this Chapter we shall be considering the ways in which the microprocessor can be involved in control of the stepping motor. Fig. 8.1, for example, shows an open-loop control

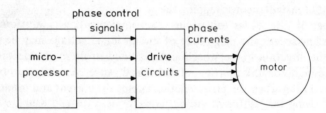

Fig. 8.1 *Software-based open-loop control*

system which would be termed 'software-intensive', because the microprocessor produces the phase control signals; the program is responsible for timing and sequencing the signals to move the motor to the required position. In complete contrast is the 'hardware-intensive' system shown in Fig. 8.2. Here the micropro-

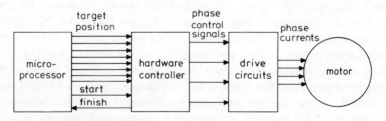

Fig. 8.2 *Hardware-based open-loop control*

cessor program merely feeds the target position information and a start command to the hardware controller, which generates the phase control signals for the motor drive circuits and a 'finish' signal for the microprocessor when the target is reached.

When choosing between software- and hardware-intensive interfaces (or possibly attempting a compromise between the extremes depicted in Figs. 8.1 and 8.2), the system designer has to consider several aspects of microprocessor and motor performance. It is immediately apparent from the preceeding discussion that a control scheme based on the software approach is likely to involve a considerable commitment of processor capacity. If this capacity is already available then the software-intensive approach can be adopted at little cost, as standard stepping motor control programs are available. However if significant additional capacity is needed then the potential benefits of software control must be balanced against the price of processor expansion. In applications involving the real-time control of several other devices the hardware-intensive approach may be the only realistic alternative, because of programming constraints.

As far as the motor is concerned the software-intensive approach makes it easier to implement more sophisticated control schemes aimed at maximising motor performance. In Section 6.4.1, for example, with microprocessor-based open-loop control it is possible to specify the duration of each step and the optimum velocity profile can be followed very closely. With a hardware-intensive approach this level of control sophistication is not usually available and the control circuit generates an approximation to the optimal operating condition, e.g. the acceleration/deceleration velocity profiles are approximated by linear ramp functions.

Finally the operating speeds of the motor and microprocessor must be reconciled. With closed-loop control motors may reach speeds of 20 000 steps per second and at these speeds a step is executed in 50 μs. Microprocessors for real-time control applications (e.g. Motorola 6800, Intel 8080) have instruction cycle times of 1–2 μs and therefore a software-based closed-loop control would be limited to 25–50 instructions per motor step at high speeds, which would restrict control to simple functions, such as step timing, step counting and phase sequencing. If the software-intensive approach is used to implement more sophisticated control schemes the motor's stepping rate is limited by the processor operating speed.

In the remainder of this Chapter the software- and hardware-intensive implementations of several control schemes are discussed in detail with particular reference to processor requirements, optimisation of motor performance and limitations imposed by microprocessor-based control.

8.2 Software vs. hardware for open-loop control

Many stepping motor systems operate with open-loop control at a constant stepping rate, which is below the start/stop rate for the most demanding load conditions. This arrangement is perfectly satisfactory for applications in which successful operation is not critically dependent on the time taken by the motor to position its load. The advantage of this system is its simplicity; a very limited number of control functions are required and the software or hardware option is available for each function. These options and their relative merits are considered in the first sub-section.

The second sub-section looks at the implementation of more sophisticated open-loop control schemes and shows that in many cases the choice of software- or hardware-based control depends on the aspects of performance which are most important in the particular application.

8.2.1   Constant rate operation

In this simple approach to open-loop control there are three basic control functions:

(*a*)   stepping rate: sets the motor speed, which must be less than the start/stop rate,

(*b*)   phase sequencing: ensures that the motor phases are excited in the order corresponding to the required direction of rotation,

(*c*)   step counting: records the number of steps taken by the motor and inhibits the step commands when the target position is attained.

Each of these control functions can be implemented with software or hardware and a complete stepping motor controller may use any combination of the software or hardware alternatives.

Software rate, sequence and count. In this system the microprocessor outputs are the phase control signals to the motor drive circuits (Fig. 8.1) and all three control functions are performed by the microprocessor program. As an example Fig. 8.3 shows the flowchart for the control of a two-phase hybrid-type stepping motor by an eight-bit microprocessor. The motor is operated with a two-phase-on excitation sequence and the program controls two bits of the Parallel Interface Adaptor (PIA), according to the required direction of rotation:

+ rotation		− rotation	
Bit 0	Bit 1	Bit 0	Bit 1
0	0	0	0
0	1	1	0
1	1	1	1
1	0	0	1
0	0	0	0

With bit 0 of the PIA controlling the polarity of phase A excitation (e.g. if bit 0 = 0 then A is excited by positive current, if bit 0 = 1 then A is excited by negative current) and bit 1 controlling phase B, the motor excitation sequences are $A+B+$, $A+B-$, $A-B-$, $A-B+$, $A+B+$, . . . or $A+B+$, $A-B+$, $A-B-$, $A+B-$, $A+B+$, . . .

The motor control routine shown in Fig. 8.3 is entered with the number of steps loaded into accumulator A (if the number of steps cannot be expressed within the

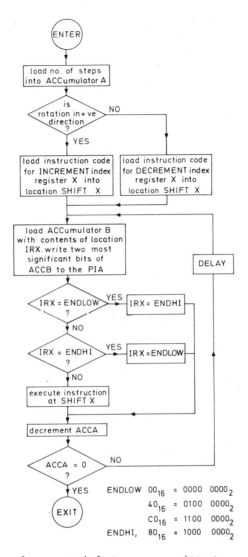

Fig. 8.3 *Flowchart for software control of rate, sequence and count*

confines of an eight-bit word — corresponding to 256 steps — the routine must be executed several times) and the index register X pointing at a location in the list between ENDLOW and ENDHI. According to the required direction of rotation the location SHIFTX is loaded with the instruction code to either increment register X (positive rotation) or decrement X (negative rotation). Accumulator B is then loaded with the contents of the store addressed by register X (indexed addressing) and the two most significant bits are written to the PIA, producing a change in phase excitation. For example, with the index register at ENDLOW+1, the number

40_{16} is placed in accumulator B and, since bit $0 = 0$ and bit $1 = 1$, the phase excitation is changed to $A+B-$.

The index register moves through the list between ENDLOW and ENDHI in the direction corresponding to the direction of motor rotation. If the motor is moving in the negative direction the excitation cycle is complete when the index register reaches location ENDLOW and the next cycle is initiated by shifting the register to point at location ENDHI. Similarly for positive rotation the index register reaches ENDHI and must be shifted back to ENDLOW. If, however, the register is not at either end of the excitation sequence list it is incremented or decremented when the instruction at SHIFTX is executed.

Accumulator A records the position of the motor relative to the target and, as the motor has moved one step due to the change in phase excitation, this count is decremented. The contents of accumulator A are then tested, a zero result indicating that the target position has been attained and the motor control routine can be exited.

If further steps are to be performed a time delay is needed before the next excitation change. During this delay the microprocessor can continue with other tasks (including, perhaps, the control of other stepping motors), the only restriction being that the time spent on these tasks must be equal to the excitation interval. On leaving the delay routine the next excitation change is initiated by loading accumulator B with the next value in the excitation list.

The main advantage of this software-based control system is the simplicity of the microprocessor/motor interfacing, which can be implemented directly by using the PIA outputs as phase control signals. Although the motor control program occupies more of the microprocessor store than the other alternatives, the storage requirements are still extremely modest. The main programming problems are likely to arise in the correct timing of the phase excitation changes; the program segment responsible for the time delay between steps must be carefully written if it is to provide accurate timing, as well as perform a useful secondary function.

Software counting, hardware timing and sequencing. This method of implementing open-loop control is shown schematically in Fig. 8.4(*a*). The stepping rate is fixed by a constant frequency clock, which is controlled by a one-bit signal from the microprocessor. Clock pulses are directed to the microprocessor, which records the motor position relative to the target, and to the excitation sequence hardware, which produces the phase control signals in the sequence dictated by the microprocessor-generated direction signal. Fig. 8.4(*b*), for example, shows a typical excitation sequence circuit, based on two $J-K$ flip-flops and four exclusive-OR gates, for two-phases-on excitation of a hybrid stepping motor. The inputs to this circuit are the clock pulses and direction signal, while the outputs are the four phase control signals $A+, A-, B+, B-$.

The microprocessor software must count the clock pulses and generate the clock start/stop and direction signals. In Fig. 8.4(*c*) the software flowchart has two branches. On entering the motor control routine the number of steps to be performed is loaded into a counter, in this case accumulator A. Accumulator B is

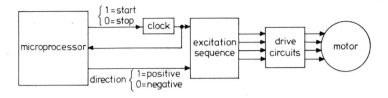

a

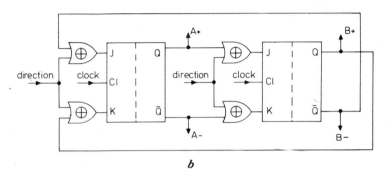

b

c

Fig. 8.4 *Software count, hardware rate and sequence control*
a   block diagram
b   hardware excitation sequence circuit
c   flowchart for microprocessor software

first cleared then, if positive rotation is required, has its most significant bit (bit 0) set. The next most significant bit (bit 1) is also set at this stage and these two bits of the accumulator are then written to the PIA lines used as the direction and clock start/stop signals. Bit 0 of accumulator B therefore appears at the output as the direction signal and bit 1 as the start/stop signal. The motor control routine can then be exited, allowing the microprocessor to perform other tasks.

When the clock receives the start signal it begins to generate pulses, which are fed to a microprocessor interrupt line, so that program execution is forced to transfer to the INTERRUPT entry of the motor control program. A clock pulse reaching the excitation sequence circuit causes an excitation change corresponding to one motor step, so the record of motor position relative to the target must be updated by decrementing accumulator A. If the accumulator is non-zero after this operation the control routine can be exited, but a zero result indicates that the target has been attained and the clock must be stopped. This is achieved by clearing accumulator B and writing its two most significant bits to the *PIA,* so that the direction and start/stop signals are both cleared.

The advantage of this software/hardware combination is that both components are quite simple. Constant frequency clock circuits controllable by a start/stop signal are available as single integrated circuits and the excitation sequence generator of Fig. 8.4(*b*) needs only two integrated circuit packages. The microprocessor software occupies a minimal amount of store and program development is not restricted by timing considerations; the microprocessor is free to perform other tasks between each motor step, provided the external clock interrupts can be tolerated.

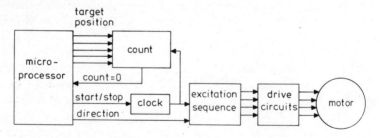

Fig. 8.5   *Hardware control of rate, sequence and count*

Hardware rate, sequence and count. In this system (Fig. 8.5) all three basic control functions are performed by hardware. At the beginning of the motor control routine the microprocessor loads the target position into a downcounter, starts the constant frequency clock and generates a direction signal for the excitation sequence circuit, but then has no further involvement until the target is reached.

The constant frequency clock produces step commands at the start/stop rate of the motor/load combination and these commands are fed to both the excitation sequence circuit and to the downcounter, which records the instantaneous motor position relative to the target. When the target is attained the counter contents are zero and this condition is signalled to the microprocessor, which then sends a stop signal to the clock. (Alternatively the 'count = 0' signal can stop the clock directly.)

The main advantage of this system is that the commitment of microprocessor capacity is minimal; in contrast to the two previous systems, the processor is not involved with each motor step. As far as hardware is concerned, this is obviously the most complex of the systems described, but the circuits can be realised with a small number of standard integrated circuit packages and purpose-built stepping motor control packages are now widely available (Sigma Instruments, 1980; Klingman, 1980). One possible disadvantage of the hardware-based system is that a relatively large number of signal lines are needed to interface the control circuits to the microprocessor, particularly for the parallel loading of the counter, e.g. for a maximum travel of 128 steps (=2^7 steps), seven signal lines are used to load the counter.

8.2.2   Ramped acceleration/deceleration

If an open-loop stepping motor control is to operate at speeds in excess of the start/ stop rate the instantaneous stepping rate must be increased over a number of steps, until the maximum speed is attained. As the target position is approached the step-ping rate must be reduced, so that the motor is running at a speed below the start/ stop rate when the target is reached. The optimum velocity profile can be found from the pull-out torque/speed characteristics of the motor and load, using the analytical techniques described in Chapter 6. Several methods of implementing open-loop control, including a microprocessor-based scheme, are introduced in that Chapter, so in this Section we can concentrate on the relative merits of software- and hardware-intensive systems for the detailed control of stepping rate.

Software-intensive approach. In this type of open-loop control the micropro-cessor generates the phase control signals (Fig. 8.1) and the processor software is responsible for phase sequencing, step counting and the timing of each excitation interval. For a ramped acceleration/deceleration control the task of excitation timing is particularly arduous and may lead to excessive storage requirements, because the length of each step during acceleration and deceleration occupies a separate storage location.

A flowchart for a typical software-based control is shown in Fig. 8.6. The length of each excitation interval is determined by a delay value, which is read from either the acceleration or deceleration look-up table. Between each excitation change the processor executes a WAIT instruction, in which the delay value is counted down to zero; high delay values produce a long delay time and therefore a low stepping rate. In the example of Fig. 8.6 typical delay values for acceleration are stored in a table beginning at location ACCST, ending at location ACCEND and containing ACCVAL values. Similarly the deceleration values occupy the DECVAL locations between DECST and DECEND.

The program begins with the target position (in steps) loaded into COUNT and a

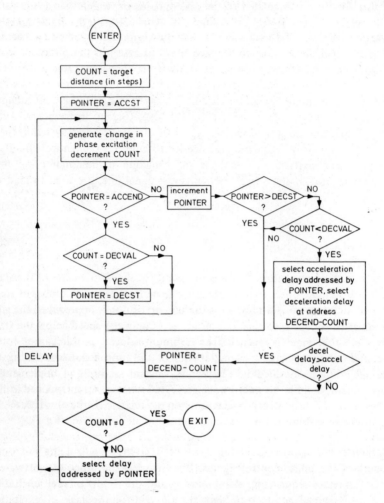

Fig. 8.6 *Flowchart for software control of ramped acceleration/deceleration*

Sample acceleration table		Sample deceleration values	
ACCST,	300	DECST,	75
	240		77
	190		80
	150		84
	115		.
	90		.
	.		.
	.		230
ACCEND,	70	DECEND,	300

ACCVAL = ACCEND − ACCST.   DECVAL = DECEND − DECST.

POINTER set at the address of the first delay value, ACCST. A change of excitation, according to the required direction of rotation, sets the motor in motion and the COUNT of steps remaining is decremented. The program proceeds by incrementing POINTER, picking up the next delay value from the acceleration table and waiting until the step interval is complete, before generating the next excitation change. However a number of simple tests must be performed to ensure that POINTER is indicating the correct delay value:

(*a*)   If the end of the acceleration table has been reached the motor continues at constant velocity with a step interval dictated by the final acceleration delay value, stored in location ACCEND. This condition can be detected by testing the equality of POINTER and ACCEND.

(*b*)   If the motor has attained its maximum stepping rate deceleration must commence when the number of steps required to decelerate is equal to the number of steps from the target. This condition is detected by comparing COUNT and DECVAL. Deceleration is initiated by setting POINTER = DECST, so that subsequent delay values are read from the look-up table.

(*c*)   The motor is unable to reach its maximum speed if the number of steps to be executed is less than the total steps required for acceleration and deceleration, i.e. the initial value of COUNT is less than ACCVAL + DECVAL.

During acceleration the number of steps to the target (COUNT) is compared to the number of deceleration values (DECVAL). As long as COUNT is greater than DECVAL the motor can continue to accelerate, but otherwise a further test is needed. The current acceleration delay is compared to the deceleration delay (in location DECEND–COUNT) which would be used if deceleration began at that position. If the deceleration delay is greater than the acceleration delay, deceleration can be initiated by setting POINTER to the appropriate location (DECEND–COUNT) in the deceleration table.

The software-intensive control scheme provides accurate and detailed timing of the step intervals during acceleration and deceleration. If the time taken to execute one cycle of delay routine is T_1 and the time occupied in changing excitation, updating and position count and moving the delay pointer is T_2, then, for a delay value d, the time between excitation changes is:

$$\text{Step interval} = dT_1 + T_2$$

which is the reciprocal of the instantaneous stepping rate:

$$\text{Stepping rate} = 1/(dT_1 + T_2)$$

T_1 and T_2 are fixed by the number of processor instruction cycles required to execute the corresponding sections of software. Typical values, for processors such as the Motorola 6800 or Intel 8080, are $T_1 = 10$ μs and $T_2 = 50$ μs and a table of stepping rates for various delays can be constructed:

Stepping rate (steps s^{-1})	Delay
99·9	996
100·0	995
100·1	994
•	•
•	•
•	•
990	96
1000	95
1010	94
•	•
•	•
•	•
4760	16
5000	15
5270	14

This table illustrates a fundamental weakness of the software-based system: the range of available stepping rates becomes coarser with increasing speed; for example, at 100 steps per second a unity change in delay value produces a 0·1% change in stepping rate, but at 5000 steps per second the same delay change gives a 5% difference in stepping rate. The stepping rate quantisation may prevent the system from approaching its pull-out rate, e.g. if the pull-out rate is 7000 steps per second, there may be insufficient torque available to accelerate between 5000 and 5270 steps per second.

As far as processor requirements are concerned, the flowchart of Fig. 8.6 assumes that the microprocessor is dedicated to the control of a single stepping motor, but the software could be adapted to the control of several motors. A greater problem is the need for a large amount of store, particularly when the stepping motor system has a heavy inertial load and many steps are needed for acceleration and deceleration. If some departure from the optimal velocity profile can be tolerated, the potentially high costs of additional store can be reduced in two ways:

(*a*)   the acceleration delay values can be used, in reverse order, for deceleration,

(*b*)   an approximate recursion formula may be devised, so that each delay value can be calculated from a limited number of previous values.

The main conclusion to be drawn from this discussion is that a software-based system gives detailed control of the velocity profile up to medium stepping rates (1000 steps per second), but may limit high-speed performance, and is therefore well-suited to applications in which acceleration/deceleration operations predominate.

Hardware-intensive approach. In a hardware-based controller the microprocessor's involvement is minimal; information on the target position and direction is passed to the control circuits (Fig. 8.2) from the processor, which may continue with other tasks until the hardware controller signals that the target has been reached. The control circuits must therefore attend to the timing and sequencing of the phase excitation, as well as recording the motor's position relative to the target.

Most control circuits are of the form shown schematically in Fig. 8.7. At the

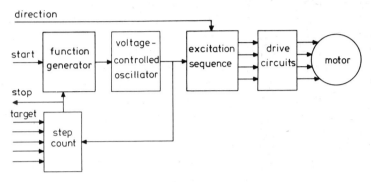

Fig. 8.7   *Hardware-based control of ramped acceleration/deceleration*

centre of this system is a voltage-controlled oscillator, which issues step commands to the phase excitation circuit. The oscillator input is provided by a function generator, which can be started by a signal from the microprocessor and stopped by a signal from the step counter. One advantage of this system over the software-intensive approach is immediately apparent: the voltage-controlled oscillator's output is continuously variable, so the maximum operating speed of the motor is no longer restricted by its ability to 'jump' between discrete stepping rates.

For optimum open-loop operation the velocity profile for acceleration is an approximate exponential function of time and for deceleration a 'reverse exponential' is needed [Fig. 8.8(*a*)]. Unfortunately this latter function is very dif-

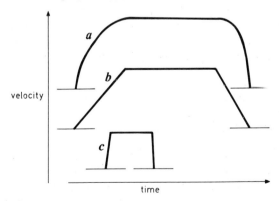

Fig. 8.8   *Open-loop control velocity profiles*
    a   optimum profile
    b   linear ramp for long travel
    c   linear ramp for short travel

ficult to generate using simple analogue circuits, so it has become common practice to approximate the velocity profile by linear ramp functions [Figs. 8.8(*b,c*)]. However this approximation imposes a constraint on the system's maximum stepping rate, which becomes a function of the ramp gradient. At any speed the maximum permissible acceleration is given by the equation:

Acceleration = (motor torque – load torque)/inertia

For example, at the pull-out rate the motor and load torques are equal, so the pullout rate is approached with zero acceleration. For a linear ramp of velocity, acceleration is constant and is limited by the maximum permissible acceleration. The maximum stepping rate must be restricted, so that the motor torque substantially exceeds the load torque at all speeds and the system can accelerate rapidly. Because motor torque decreases with increasing speed, a low stepping rate limit implies a high motor torque at all speeds and therefore fast acceleration, so the limiting speed is attained in a short time [Fig. 8.8(*c*)]. Conversely a high stepping rate limit leads to a low acceleration rate [Fig. 8.8(*b*)]. The optimum ramp gradient and stepping rate limit depend on the number of steps to be executed (see following example) and, in principle, the function generator characteristics could be made to depend on the target position information.

A number of general-purpose stepping motor controllers – often in the form of MSI integrated circuits – are now available and provide various levels of control sophistication (Sigma Instruments, 1980). An interesting recent development is the programmable stepping motor control circuit (Klingman, 1980), which provides a useful compromise between dedicated microprocessor control, with its high cost, limited speed range but good acceleration/deceleration characteristics, and hardware-intensive control, with low cost, wide speed range but poor acceleration/deceleration characteristics.

Example

A 15-degree stepping motor has the approximate torque/speed characteristic shown in Fig. 8.9 and is used to drive a load inertia of 10^{-4} kg m^2. If the hardware

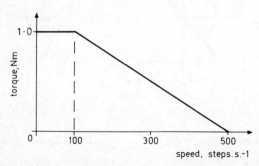

Fig. 8.9   *Pull-out torque/speed characteristic*

open-loop controller provides linear ramped acceleration/deceleration, find the optimum stepping rate limit and corresponding ramp gradient to minimise the time taken to move 50 steps.

For any stepping rate limit the motor torque can be read-off the torque/speed characteristic and the corresponding rate of acceleration calculated from:

$$\text{Acceleration} = \text{motor torque/inertia}$$

Stepping rate limit (steps s^{-1})	Motor torque (Nm)	Acceleration	
		(rads s^{-2})	(steps s^{-2})
300	0·50	5000	19 000
350	0·37	3700	14 000
400	0·25	2500	9 500
450	0·12	1200	4 800

The time taken to accelerate to the maximum speed is:

$$\text{Time} = \text{speed/acceleration}$$

and the distance travelled during acceleration is:

$$\text{Distance} = \text{speed} \times \text{time}/2$$

In general the motor is able to decelerate faster than it can accelerate, but if a single ramp is to control both acceleration and deceleration the ramp rate must be matched to the acceleration rate. The times and distances involved in deceleration are then the same as in acceleration.

For a trial distance of 50 steps the number of steps executed at the maximum speed is obtained by subtracting the steps executed during acceleration and deceleration. The following table can then be constructed:

Stepping rate limit (steps s^{-1})	Acceleration + deceleration		Maximum speed		Total Time
	Time (ms)	Distance (steps)	Distance (steps)	Time (ms)	(ms)
300	31	4·7	45·3	151	182
350	50	8·7	41·3	118	168
400	84	16·8	33·2	83	167
450	188	42·3	7·7	17	205

So, for a travel of 50 steps, the time taken is minimised when the stepping rate limit is set between 350 and 400 steps per second, with the acceleration/deceleration rate between 14 000 and 9500 steps per second^2.

8.3   Microprocessor-based closed-loop control

In a closed-loop stepping motor system there are five basic control functions:

(a)   step count: records the number of steps executed,

(b)   phase sequence: excites the motor phases in the sequence appropriate to the direction of rotation,

(c)   position detection: produces a signal pulse as each step is completed,

(d)   ignition advance: varies the switching angle with motor speed,

(e)   deceleration initiation: detects the proximity of the target position and signals the phase sequence generator.

The first two control functions – step count and phase sequence – are common to open-loop systems and, as we have already seen, may be implemented with software or hardware. Optical and waveform methods of position detection are discussed in Chapter 7 and in the latter case the possibilities of a system based on software analysis of current waveforms cannot be ignored. However the operating speeds of current microprocessors would limit the useful speed range of any such system; at motor speeds of 10 000 steps per second, for example, the step interval is 100 μs and in this time around 70 instruction cycles could be executed, hardly sufficient for reliable analysis of the waveform data.

Microprocessor technology can be usefully employed for the final two control functions – ignition advance and deceleration initiation – so these topics are discussed separately in the following sub-sections.

8.3.1   Control of switching angle

In a closed-loop control scheme with continuously-variable switching angle the motor is able to develop its pull-out torque at all speeds and therefore the system performance is maximised. To perform this function the controller requires information about the instantaneous speed of the motor and must then generate the switching angle appropriate to that speed.

Fig. 8.10(a) shows the block diagram of a typical system, in which a dedicated microprocessor receives position signals from the motor and transmits step commands to the excitation sequence circuits. The processor software – Fig. 8.10(b) – begins by issuing a step command to set the stepping motor in motion. A counter (COUNT) is incremented as the program cycles around the short loop. During this first step the variable FIRE is zero, so no step commands are issued before the first position detector pulse is received. The position detector is orientated relative to the rotor so as to generate signals at the optimum switching angle for low motor speeds and therefore, at the end of the first step, a step command is issued immediately.

During all subsequent steps the switching angle is controlled by the software, which tests for equality of COUNT and FIRE. The time between the first step command and the position detector pulse signalling completion of the first step is

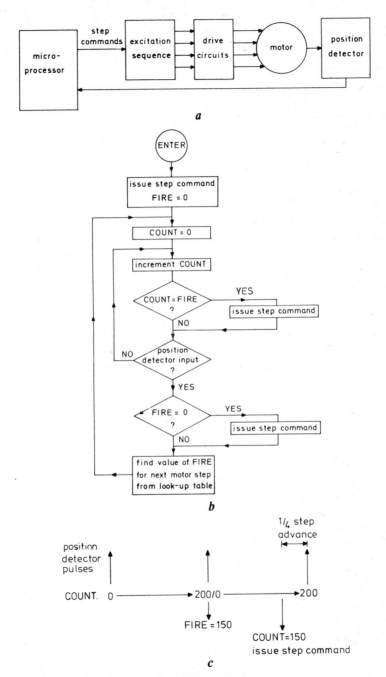

Fig. 8.10   *Microprocessor-based control of switching angle*
a   block diagram
b   flowchart for software
c   timing diagram for 1/4 step ignition advance

proportional to the value of COUNT, which is inversely proportional to the average motor speed over the step. A look-up table is used to find the switching angle – dictated by the value of FIRE – appropriate to this speed. For example, if COUNT is 200 when the motor is running at 500 steps per second and 1/4 step of ignition advance is required, then FIRE is set to 150. During the next step COUNT is incremented until it is equal to FIRE (after 3/4 step) and a step command is issued, 1/4 step before the next position detector signal, as in the timing diagram, Fig. 8.10(c).

In the basic system described here the switching angle effective for one step depends on the average motor speed over the previous step. This is perfectly satisfactory for steady-state operation, but a more sophisticated arrangement is needed if the system is to contend with rapid speed changes, such as occur during acceleration from rest with light inertial loading. Radelescu and Stoia (1979), for instance, describe a microprocessor-based closed-loop control in which the switching angle is calculated by an algorithm involving the two previous step periods, enabling the rate of change of speed to be taken into account.

8.3.2   Deceleration initiation and adaptive control

If a closed-loop stepping motor control is to minimise the time spent in moving a load to a target position a velocity profile of the form shown in Fig. 8.11(a) must

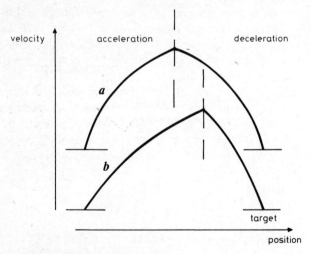

Fig. 8.11   *Closed-loop control velocity profiles*
a   acceleration = deceleration
b   acceleration < deceleration (high load torque)

be used. The motor accelerates rapidly to high speeds and deceleration is initiated at a point where the maximum rate of deceleration enables the load to reach the target position with zero speed. In this example the load torque is negligible, so the rates of acceleration and deceleration are equal and deceleration is initiated at a point exactly halfway to the target. Any load torque opposes acceleration and aids deceleration of the system, so in the case of Fig. 8.11(*b*), where the load torque is high, deceleration is initiated closer to the target.

In setting up a closed-loop stepping motor control the choice of appropriate points for deceleration initiation can be very difficult, because the load torque characteristics, which are usually a function of speed, may be ill-defined or may be expected to vary over the life expectancy of the system. However a microprocessor-based control system can be 'taught' to initiate deceleration at the appropriate point, if it is programmed to learn from its previous attempts at producing an optimal trajectory. Kenjo and Takahashi (1980) describe a printer character-wheel control which produces movements of up to 200 character positions in minimum time. The system learns the optimum point for deceleration initiation for each movement by making several attempts, which give successively better results. After installation any changes in load conditions, e.g. due to bearing wear, are automatically corrected by updating the stored deceleration initiation points for each movement.

This learning process, which is presently based on the results of a series of trial-and-error experiments, is capable of further refinement. For a system with a known load inertia and motor torque/speed characteristic, variations in the velocity profile during acceleration are indicative of the changes in the load torque/speed characteristic. If the closed-loop control microprocessor receives signals from the position detector the acceleration velocity profile can be constructed and it is then possible to characterise the load and, given sufficient processor capacity, anticipate the deceleration trajectory. An optimal velocity profile can be provided for every movement, with the deceleration initiation point relative to the target position being adapted to the instantaneous load conditions.

Stepping motors and microprocessors have evolved separately, but the properties of the two devices are complementary and their combination provides a powerful control mechanism. The systems described here represent the early attempts at combining the capabilities of stepping motors and microprocessors; a great deal of development will be needed before the full potential of the combination is realised.

Chapter 9

Appendix: pull-out torque/speed characteristics of bifilar-wound motors

The theory of torque production in the hybrid stepping motor, presented in Section 5.2, requires some modification if the motor is bifilar-wound. Each phase is split into two bifilar windings, which have equal numbers of turns and are located in the same stator poles, but are wound in opposite senses. If phase A is split into bifilar windings $A+$ and $A-$, then the magnet flux linked with each winding is:

$$\psi_{A+} = \psi_M \sin(p\theta)$$

$$\psi_{A-} = -\psi_M \sin(p\theta)$$

(9.1)

where ψ_m is the peak magnet flux linking *one* bifilar winding.

A unipolar drive circuit is used to excite the bifilar windings, so the dc and fundamental components of applied voltage must be considered:

$$v_{A+} = V_0 + V_1 \cos(\omega t)$$

$$v_{A-} = V_0 - V_1 \cos(\omega t).$$

(9.2)

Note that the fundamental voltage components are in anti-phase, as are the fundamental current components:

$$i_{A+} = I_0 + I_1 \cos(\omega t - \delta - \alpha)$$

$$i_{A-} = I_0 - I_1 \cos(\omega t - \delta - \alpha)$$

(9.3)

If each bifilar winding circuit has a total resistance (including forcing) of R and a self-inductance $L/2$, the mutual inductance between bifilar windings must also be $L/2$ (Section 2.4) and the voltage equation for winding $A+$ becomes:

$$v_{A+} = Ri_{A+} + (L/2)\, di_{A+}/dt + (L/2)\, di_{A-}/dt + d\psi_{A+}/dt$$

and substituting from Eqns. (9.1) and (9.3):

$$V_0 + V_1 \cos(\omega t) = RI_0 + RI_1 \cos(\omega t - \delta - \alpha) - (L/2)I_1 \sin(\omega t - \delta - \alpha)$$

$$- \omega(L/2)I_1 \sin(\omega t - \delta - \alpha) + \omega\psi_M \cos(\omega t - \delta)$$

Separating the d.c. and fundamental voltage components in this expression:

$$V_0 = RI_0 \tag{9.4}$$

and

$$V_1 \cos t(\omega t) = RI_1 \cos(\omega t - \delta - \alpha) - \omega L I_1 \sin(\omega t - \delta - \alpha) + \omega\psi_M \cos(\omega t - \delta)$$

$$\tag{9.5}$$

Eqn. (9.5) is identical in form to Eqn. (5.9) for one phase of the conventional hybrid motor, so the circuit models are identical and we can proceed directly to the expression for pull-out torque. Eqn. (5.14) gives the pull-out torque for a two-phase hybrid motor, but the bifilar-wound motor has four windings, so the pull-out torque expression must be multiplied by a factor of two:

$$\text{Pull-out torque} = \frac{2_p \, \psi_M \, V_1}{(R^2 + \omega^2 L^2)^{1/2}} - \frac{2p \, \omega \, \psi_M^2 \, R}{(R^2 + \omega^2 L^2)} \tag{9.6}$$

where

V_1 = fundamental component of voltage applied to *one* bifilar winding

L = 2 x (self-inductance of one bifilar winding)

R = resistance of a bifilar winding circuit

ψ_M= peak magnet flux linked with one bifilar winding

p = number of rotor teeth

ω = $(\pi/2)$ x stepping rate

References

ACARNLEY, P. P., HUGHES, A., and LAWRENSON, P. J. (1979): 'Torque and speed capabilities of multi-stack VR motors', Proceedings of the international conference on stepping motors and systems, University of Leeds

ACARNLEY, P. P., and HUGHES, A. (1981): 'Predicting the pullout torque/speed curve of variable-reluctance stepping motors', *Proc. IEE,* **128,** Part B, No. 2

BAKHUIZEN, A. J. C. (1973): 'A contribution to the development of stepping motors', Ph.D. Thesis, Technological University, Eindhoven, Holland

BAKHUIZEN, A. J. C. (1976): 'Considerations on improving the ratio of torque to inertia', Proceedings of the international conference on stepping motors and systems, University of Leeds

BAKHUIZEN, A. J. C. (1979): 'On self-synchronisation of stepping motors', Proceedings of the international conference on stepping motors and systems, University of Leeds

BARNARD, W. R., and LLOYD, J. J. (1980): 'The changing requirements of teaching electrical power and machines', Fifteenth universities power engineering conference, University of Leicester

BELING, T. E. (1978): 'A ramper circuit for step motor control', Proceedings of the eighth annual symposium on incremental motion control systems and devices, University of Illinois

BELL, R., LOWTH, A. C., and SHELLEY, R. B. (1970): 'The applications of stepping motors to machine tools' (The Machinery Publishing Company)

CASSAT, A. (1977): 'High-performance active-suppression driver for variable-reluctance step motors', Proceedings of the sixth annual symposium on incremental motion control systems and devices, University of Illinois

DAVIES, E. (1980): 'Using VMOS in microprocessor stepping motor control', Electronic Engineering

ERTAN, H. B., HUGHES, A., and LAWRENSON, P. J. (1980): 'Efficient numerical method for predicting the torque displacement curve of saturated VR stepping motors', *Proc. IEE,* **127,** Part B, No. 4

FINCH, J. W., and HARRIS, M. R. (1979): 'Linear stepping motors: an assessment of performance', Proceedings of the international conference on stepping motors and systems, University of Leeds

FITZGERALD, A. E., and KINGSLEY, C. (1952) 'Electric machinery' (McGraw-Hill).

FRUS, J. R., and KUO, B. C. (1976): 'Closed-loop control of step motors using waveform detection', Proceedings of the international conference on stepping motors and systems, University of Leeds

GERBER, H. S. (1975): U.S. Patent 3,445,741

GUPTA, R. K., and MATHUR, R. M. (1976): 'Transient performance of variable-reluctance stepping motors', IEE Conference Publication No.136

HARRIS, M. R., ANDJARGHOLI, V., LAWRENSON, P. J., HUGHES, A., and ERTAN, B. (1977): 'Unifying approach to the static torque of stepping-motor structures', *Proc. IEE*, **122**, No. 8

HUGHES, A., and LAWRENSON, P.J. (1975): 'Electromagnetic damping in stepping motors'. *Proc. IEE*, **122**, No. 8

HUGHES, A., LAWRENSON, P. J., and DAVIES, T. S. (1976): 'Pull-out torque characteristics of hybrid stepping motors', Proceedings of the international conference on stepping motors and systems, University of Leeds

JUFER, M. (1976): 'Self-synchronisation of stepping motors', Proceedings of the international conference on stepping motors and systems, University of Leeds

KENJO, T., and TAKAHASHI, H. (1980): 'Microprocessor controlled self-optimization drive of a step motor', Proceedings of the ninth annual symposium on incremental motion control systems and devices, University of Illinois

KENT, A. J. (1973): 'An investigation into the use of inertia dampers on step motors', Proceedings of the second annual symposium on incremental motion control systems and devices, University of Illinois

KLINGMAN, E. E. (1980): 'A stored program stepper motor controller chip', Proceedings of the ninth annual symposium on incremental motion control systems and devices, University of Illinois

KORDIK, K. S. (1975): 'Step motor inductance measurements', Proceedings of the fourth annual symposium on incremental motion control systems and devices, University of Illinois

KUO, B. C. (1974): 'Theory and applications of step motors' (West Publishing Company)

KUO, B. C., and CASSAT, A. (1977): 'On current detection in variable-reluctance step motors', Proceedings of the sixth annual symposium on incremental motion control systems and devices, University of Illinois

LAFRENIERE, B. C. (1979): 'Software for stepping motors', *Machine Design* **51**, Part 9

LAJOIE, P. A. (1973) 'The incremental encoder − an optoelectronic commutator', Proceedings of the second annual symposium on incremental motion control systems and devices, University of Illinois

LANGLEY, L. W., and KIDD, H. K. (1979): 'Closed-loop operation of a linear stepping motor under microprocessor control', Proceedings of the international conference on stepping motors and systems, University of Leeds

LAWRENSON, P. J., and KINGHAM, I. E. (1974): 'Design of viscously-coupled inertial dampers', Proceedings of the international conference on stepping motors and systems, University of Leeds

LAWRENSON, P. J., and KINGHAM, I. E. (1977): 'Resonance effects in stepping motors', *Proc. IEE*, **124**, No. 5

LAWRENSON, P. J., HUGHES, A., and ACARNLEY, P. P. (1977): 'Improvement and prediction of open-loop starting/stopping rates of stepping motors', *Proc. IEE*, **124**, No. 2

MAGINOT, J., and OLIVER, W. (1974): 'Step motor drive circuitry and open-loop control', Proceedings of the third annual symposium on incremental motion control systems and devices, University of Illinois

PICKUP, I. E. D., and TIPPING, D. (1976): 'Prediction of pull-in rate and settling time characteristics of a variable-reluctance stepping motor and effect of stator-damping coils on these characteristics', *Proc. IEE*, **124**, No. 2

PRITCHARD, E. K. (1976) 'Mini-stepping motor drives', Proceedings of the fifth annual symposium on incremental motion control systems and devices, University of Illinois

RADELESCU, M. M., and STOIA, D. (1979): 'Microprocessor closed-loop stepping motor control', Proceedings of the international conference on stepping motors and systems, University of Leeds

ROOY, G., GOELDEL, G., and ABIGNOLI, M. (1979): 'An original tutorial unit of step motors', Proceedings of the eighth annual conference on incremental motion control systems and devices, University of Illinois

ROTERS, G. C. (1941): 'Electromagnetic devices' (Wiley)

RUSSELL, A. P., and LEENHOUTS, A. C. (1980): 'An application-orientated approach to the prediction of pull-out torque/speed curves for permanent magnet stepping motors', Proceedings of the ninth annual symposium on incremental motion control systems and devices, University of Illinois

SIGMA INSTRUMENTS (1972): 'Sigma stepping motor handbook' (Sigma Instruments Inc.)

SIGMA INSTRUMENTS (1980): 'Sigma stepping motor control chip set' (Sigma Instruments Inc.)

SINGH, G., LEENHOUTS, A. C., and KAPLAN, M. (1978): 'Accuracy considerations in step motor systems', Proceedings of the eighth annual symposium on incremental motion control systems and devices, University of Illinois

TAL, J. (1973): 'The optimal design of incremental motion servo systems', Proceedings of the second annual symposium on incremental motion control systems and devices, University of Illinois

TAL, J., and KONECNY, K. (1980): 'Step motor damping by phase biasing', Proceedings of the ninth annual symposium on incremental motion control systems and devices, University of Illinois

TURNER, W. W. (1978): 'Introducing step motors to undergraduates', Proceedings of the seventh annual symposium on incremental motion control systems and devices, University of Illinois

WARD, P.A., and LAWRENSON, P.J. (1977): 'Backlash, resonance and instability in stepping motors', Proceedings of the sixth annual symposium on incremental motion control systems and devices, University of Illinois

Index